Climate
and
Environmental
Systems

BLUNDELL'S SCHOOL

D. C. Money

NAME	TERM ISSUED	FORM
Gareth Evans	90/1	L6B₁
Alex Heffron	90/2	L^{VI}B²
Sarah Franklin	98/1	26d

MASLANDS

Cover photograph: Remote sensing gives a comprehensive, yet detailed view of the pattern and relative energy of air flows, and of condensation at different levels in the atmosphere. It indicates surface conditions; here showing the contrasts between the Greenland ice-caps and coastal fringes, and the cloud-free areas of north-west Europe.

Cloud formations show the fronts associated with a deep depression. Its centre is indicated by cloud swirls as it approaches Iceland, hidden by thicker cloud at the occluded front north of Scotland. Thickening cloud of a warm front approaches the British Isles, with variable cloud and clear patches in the warm sector. Behind the cold front, cells of cloud give a flecked pattern in the cold air drawn in from the north.

The author and publisher would like to thank the following for permission to use photographs:
Bureau of Meteorology, Melbourne, Figures 4.38, 4.39
European Space Agency, Figure 5.21
National Remote Sensing Centre, Plates 20, 21 and 22
Dr Paul Curran, Plate 11

We would also like to thank the organisations and individuals who gave permission for the use of data:
The Controller of Her Majesty's Stationery Office, Figures 4.17 A–D; the Meteorological Bureau, Perth, Figure 4.41; the National Remote Sensing Centre, Figure 3.17; and C. W. Thornthwaite Associates, Figures 10.16, 10.22, 10.23 and 10.24.

Published in 1988 by
UNWIN HYMAN LIMITED
15–17 Broadwick Street
London W1V 1FP

Reprinted 1989

British Library Cataloguing in Publication Data

Money, D.C. (David Charles)
Climate and environmental systems.
1. Environment. Effects of climate
I. Title
333.7

ISBN 0-7135-2844-3

Designed by Bob Wright
Illustrations by Jillian Luff and Kevin Maddison
Cover photograph reproduced by permission of The Telegraph Colour Library
Cover design by Peter and Alex Tucker

Typeset by Nene Phototypesetters Ltd, Northampton
Printed and bound in Great Britain by
Butler & Tanner Ltd, Frome and London

CONTENTS

REFERENCES IN TEXT

Superior numbers, used throughout the book, relate to the following references:

1 MEEHL, G.A. (1987) The tropics and their role in the global climate system; *Geographical Journal,* vol.153, p.21
2 TAYLOR, J.A. (1960) Methods of soil study; *Geography,* vol.45, p.52
3 IDSO, S.B. (1984) CO_2, Climate, and Consensus Science (review); *Geographical Journal,* vol.150, p.374
WARRICK, R.A. (1988) CO_2, Climatic Change and Agriculture; *Geographical Journal,* vol.154, p.221
4 MUSK, L.F. (1983) Outlook Changeable (The Sahel); *Geographical Magazine,* vol. 50, p.533
5 SIOLI, H. (1985) The effects of deforestation in Amazonia; *Geographical Journal,* vol.151, p.197
6 Report: National Academy of Science (NRSC Washington) (1975) Understanding Climatic Change
7 CHANDLER, T.J. (1962) London's Urban Climate; *Geographical Journal,* vol.128, p.279
8 CURRAN, P.J. (1983) Estimating Green LAI from multispectral aerial photography; *Photogrammetric Engineering and Remote Sensing*, Vol.49, p.1709
9 National Remote Sensing Centre Fact Sheet A04
10 NRSC Fact Sheet (A01), data: the Macaulay Institute for Soil Research

UNITS

M	mega-	one million	10^6
k	kilo-	one thousand	10^3
h	hecto-	one hundred	10^2
da	deka-	ten	10^1
d	deci-	one tenth	10^{-1}
c	centi-	one hundredth	10^{-2}
m	milli-	one thousandth	10^{-3}
μ	micro-	one millionth	10^{-6}

length
  $1 \text{ km} = 10^3 \text{ m}$
  $1 \text{ μm} = 10^{-6} \text{ m}$

area
  1 hectare (ha) = 10 000 m^2 (2.47 acres)

mass
  $1 \text{ tonne} = 10^3 \text{ kg}$

force
  1 Newton (N) accelerates a mass of 1 kg by 1 m sec^{-2}

pressure
  $1 \text{ mb} = 10^2 \text{ N m}^{-2}$

energy: work
  1 calorie (cal) = 4.186 joules (J)

power
  1 watt (W) = 1 joule sec^{-1}
  1 langley min^{-1} = 1 cal cm^{-2} min^{-1}

latent heat of fusion
  ice/water $L_f = 3.33 \times 10^5$ J kg^{-1} at 0°C

latent heat of vaporisation
  liquid/water vapour $L_v = 2.48 \times 10^6$ J kg^{-1} at 10°C

TEMPERATURE CONVERSION

°C	−70	−60	−50	−40	−30	−20	−15	−10	−5	0	1	2	3	4	5	6	7	8	9	10
°F	−94	−76	−58	−40	−22	−4	5	14	23	32	34	36	37	39	41	43	45	46	48	50

°C	11	12	13	14	15	16	17	18	19	20	21	22	23	24	25	26	27	28	29	30
°F	52	54	55	57	59	61	63	64	66	68	70	72	73	75	77	79	81	82	84	86

°C	31	32	33	34	35	36	37	38	39	40	41	42	43	44	45	46	47	48	49	50
°F	88	90	91	93	95	97	99	100	102	104	106	108	109	111	113	115	117	118	120	122

(°F equivalent to nearest whole number)

INTRODUCTION

In this book we look in some detail at the atmosphere, biosphere, and soils. They are at first considered separately in order to examine the processes involved in creating their main characteristics. But they do not exist or act in isolation. Climatic elements, living organisms, and inorganic materials are continuously interacting as part of our environment. We, too, as people, become involved in these dynamic systems.

The concept of an **ecosystem** is a useful one. It implies that plant and animal communities, and all that makes up the non-living world about them, form unified systems. These can be as small as a garden pond or as extensive as a tropical rainforest. The plants and animals, the atmosphere, the rocks and other non-living things all interact, directly or indirectly, and each is part of the environment of the others. The system is maintained by a through-flow of energy and matter, which also circulates between the various components within the system.

The interactions and feedbacks between components of the lithosphere, hydrosphere, atmosphere, and biosphere are so complex that we can only begin to understand the relationships between them by simplification. So we define and isolate open systems and sub-systems on a scale suitable for practical observation and examination. Fig. 1 shows a system through which mass and energy continuously pass, with various transfers within the system itself. Climatic elements are seen as external control variables.

We can investigate such relationships at different scales – at the macro (large), meso (medium), or micro (small) levels. For instance, a rainforest may occupy a particular area mainly in response to favourable mean monthly air temperatures and moisture content throughout the year: these are macro-climatic features. But within the forest there are considerable differences between the properties of air at tree-top level, air at the surface, air a metre or so above the surface, and air in soil

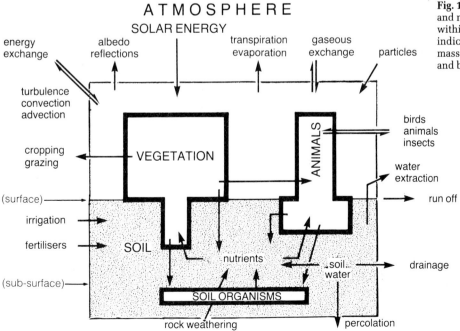

Fig. 1 The flows of energy and matter through and within a biome, with an indication of the relative mass of living matter above and below the surface.

spaces beneath the surface. So the structure of the leaves, stem, and roots of a particular plant responds to quite different micro-climates, and the balance between plant and animal species will be different at each level.

Fig. 1 shows, diagrammatically, the proportion of life-forms in a chosen system. This proportion depends on competition between the numerous plant and animal species for nutrients, light and living space. The nature of the external environmental variables will affect this momentary balance – for the system is always readjusting itself. In various parts of the world, however, the external variables have acted over long periods of time with relatively little variation on the meso scale. Occasionally, weather conditions have unusual influences, but overall there are recognisable seasonal characteristics of temperature, moisture, light conditions, and air movements. There *are* discernable patterns of climate, flora and fauna over the earth's surface. We can identify broad **biomes** related to certain macro-climatic controls (Fig. 2).

Such patterns can be identified, to some extent, in the distribution of soil types. Soil materials are derived from a variety of rocks, but responses to local climatic conditions and the associated vegetation lead to zonal soil patterns which change only slowly over long periods of time.

Nevertheless, variations in slopes, drainage and micro-climatic features can create contrasting soil types within a very small area.

Because we, as people, are part of many ecosystems, we can disturb the finely balanced relationships that exist within them. We have long been doing so, and our numbers are rapidly increasing. Therefore, although a vegetation map, such as Fig. 10.1, is a useful simplification for study purposes, it lacks reality.

Our soaring population and rapid advances in high technology are having increasingly alarming effects on local and global systems. The results are often immediately apparent. Atmospheric modifications are less easy to observe, though the effects of interference can become visible, as with acid rain.

Fortunately, we are improving **our ability to monitor changes** in the earth–atmosphere system with greater precision and in considerable detail, especially through satellite surveillance. Some of the more advanced techniques are available for use by student groups, and data from space observations and measurements can be acquired for study purposes. It is appropriate, therefore, that this book concludes by examining the complexities of rural–urban evironmental systems, and our increasing ability to monitor what is happening within them.

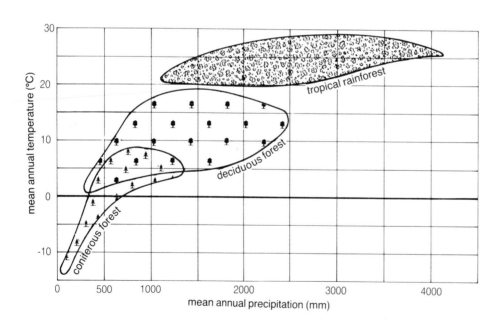

Fig. 2 The macro-climatic requirements of three major forest biomes.

Part One

The Atmosphere and Ecosystems

1

EARTH – ATMOSPHERE

1.1 The earth–atmosphere system

The earth and its atmosphere (E–A) can be seen as a closed system exchanging energy with space.

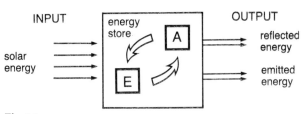

Fig. 1.1

All bodies possessing energy emit radiation. When a body's temperature rises its radiant output increases, and so does the proportion of energy of shorter wavelength.

The E–A receives energy from the very hot sun in the form of radiation (**insolation**), mostly with wavelengths varying from ultra-violet, through visible light, to infra-red. Some is of very short wavelength, such as gamma rays and X-rays, and there are small amounts of long-wave radiation, from the order of radio waves to those measured in tens of metres.

The earth emits long-wave radiation. Most of the energy in the range of 3.0–100 μm is absorbed by atmospheric gases and vapours. But there is a gap in the absorptive capacity of the gases which

allows radiation of 8–11 μm wavelength to pass through a cloudless sky and escape into space. However, this **radiation window** is partially closed by clouds and atmospheric pollutants.

Thus, much of the solar energy reaching the outer atmosphere penetrates to the surface, where it is reflected or absorbed; while the gases and vapours of the lower atmosphere hinder the loss of long-wave energy from the earth to space. These energy exchanges, and the annual E–A radiation balance, are summarised in Fig. 2.1.

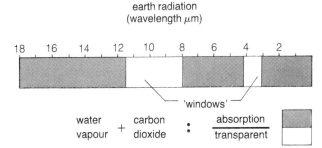

Fig. 1.3 The gaps in the ability to absorb long-wave radiation may be closed by water droplets (clouds).

At any moment half the E–A system is receiving solar energy, as the earth rotates. The amount of direct insolation received varies with latitude (p.9). Yet, on an annual scale, the system becomes neither hotter nor cooler. Energy is re-distributed

ELECTRO-MAGNETIC SPECTRUM

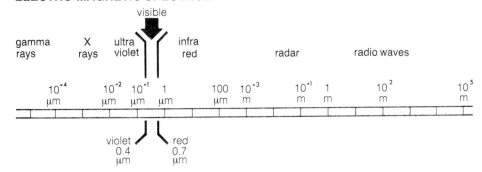

Fig. 1.2 The energy spectrum from the highly penetrative short-wave gamma-rays to those used for radio transmission.

within the system between tropical and polar latitudes by atmospheric movements and oceanic circulation. We therefore need to consider the exact nature of the atmosphere and how the energy is transferred in atmospheric movements.

1.2   The composition of the atmosphere

The earth's atmosphere is a mixture of gases and vapours. It is in ceaseless motion, but retained by the force of the earth's gravity. It is densest at sea level. About half its mass is below 6000 m, and all but 3 per cent is within 30 km of the earth's surface.

Dry surface air consists mainly of nitrogen, about 78 per cent, and oxygen, 21 per cent by volume. Among the smaller quantities of other gases, carbon dioxide, some 0.03 per cent, and the varying amounts of water vapour, are extremely important to ecosystems.

At high levels ozone (O_3) is created as molecules of oxygen (O_2) are acted on by ultra-violet radiation. This ozone concentration itself absorbs ultra-violet radiation, and gives some protection to life-forms which are harmed by excess energy of this wavelength.

We are principally affected by the lower, denser part of the atmosphere, known as the **troposphere**. Our familiar weather phenomena, including the massive towering thunderclouds, are confined to the troposphere, even though their cause may not be. Here varying amounts of tiny particles, such as dust, smoke, and salt crystals, have high local concentrations and affect the weather conditions in a number of ways.

In the troposphere there is usually a fairly rapid fall in density and temperature with altitude, to a height of some 8000 m near the poles and 17 000 m in the tropics. Here, where the temperature is about −50°C, is the **tropopause**. Above this upper boundary of the troposphere the temperature begins to rise again. The actual height of the tropopause varies with seasonal conditions.

Above the tropopause is the dust-free, cloudless **stratosphere**, almost unaffected by the turbulent upsurges of air in the troposphere. Here the ozone concentration absorbs solar radiation, causing a rise in temperature. This concentration is greatest over the polar regions, where the stratosphere becomes much warmer in summer. In winter, however, the lack of insolation means that the upper air over the polar regions is very cold. The seasonal contrasts in upper air temperature and density in the higher and lower latitudes creates strong horizontal winds at those levels.

About 50 km up the rise in temperature ceases. Above this **stratopause** it falls again to about −90°C in the **mesosphere**. Then at about 90–100 km it rises once more to the high temperatures of the outer atmosphere. The density of gases is now

very low indeed, but short-wave solar radiation separates electrons from oxygen atoms. These, with the resulting positive ions, create a zone, the **ionosphere**, which reflects radio waves. Above 350 km the atmosphere is so rarefied that molecular collisions are infrequent. The molecules circle the earth like small satellites.

At this altitude man-made geostationary satellites orbit the earth (p.166), and observe conditions at the surface and in the troposphere. A satellite travels horizontally at a speed sufficient for the outward-propelling centrifugal acceleration to match the gravitational pull at that height. The period in which it completes one earth orbit

Fig. 1.4   Notice that in the atmosphere generally the temperature rises up to some 53 km altitude and then falls to the absolute minimum about 87 km up. But in winter over the polar regions the atmospheric temperature above the troposphere continues to decline with altitude (dotted line) to a minimum approaching −90°C. Above the stratosphere, where photo-chemical effects in the thin air create a high proportion of ions, the temperature rises again.

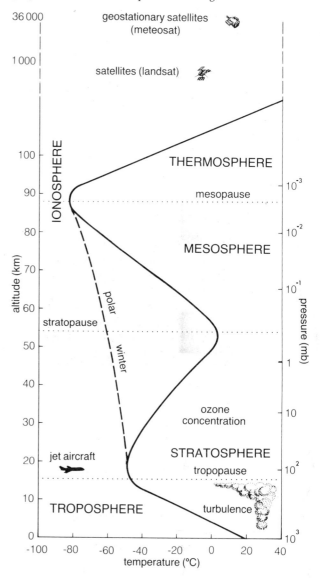

increases with its altitude. At 35 850 km, when the orbital speed is 3.35 km per sec, the period is 24 hours. So such a geosynchronous satellite circling the equator at this height would appear stationary in the sky.

We are concerned mainly with climatic conditions in the troposphere, and so we start by considering interrelated atmospheric elements: pressure gradients and resulting air flows; solar and surface energy inputs and the air temperatures which create pressure gradients at various levels in the atmosphere; and the moisture content, which is part of the earth's hydrocycle and an important source of atmospheric energy, released on condensation.

1.3 Air pressure at various altitudes

At mean sea level the normal air pressure is 1013 mb. But meteorologists refer to sea level as the **1000 mb surface**.

At any location the air pressure depends on the weight of the overlying atmosphere, and so varies with air temperature and density. Close to the surface, near sea level, the pressure falls with altitude by about 1 mb per 10 metres. Above this, as the air becomes less dense, the decrease is less rapid. The mean level of the 500 mb surface is about 5500 m and that of the 300 mb surface about 9000 m.

A **pressure gradient force** causes air to move

Fig. 1.5 Air pressure and surface gradient wind.

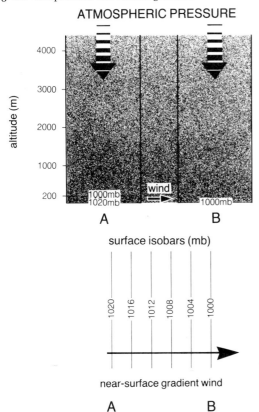

Fig. 1.6 An upper gradient wind subject to Coriolis deflection, as in Fig. 1.7.

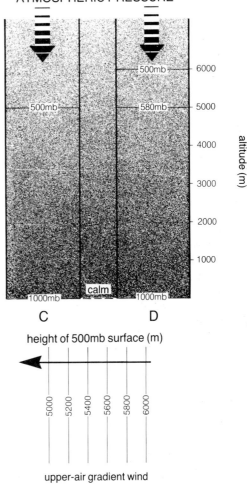

away from an area of high pressure towards places where the air pressure is lower. Fig. 1.5 shows air flows (wind) responding to pressure differences near sea level. The wind is shown blowing down a gradient represented by the isobars. Fig. 1.6 shows the contours of the height of the 500 mb surface in the atmosphere above places C and D, where the surface air is calm. You can see that an upper air gradient wind is established. On calm days at the surface, cloud movements may indicate air flows at high altitudes.

1.4 The influence of earth's rotation (the Coriolis force)

Air flowing above the rotating earth is subject to apparent deflection: to the right of the line of motion in the northern hemisphere, to the left in the southern hemisphere (**Ferrel's Law**). The deflection is greatest at the poles and diminishes to zero at the equator.

As the air moves northward or southward

across the surface, the meridians, in a sense, rotate from under it. To an earth-bound observer the wind apparently changes direction. We can allow for this by assuming that a force F acts at right angles to the wind direction (**the Coriolis force**). Where V is the wind velocity, θ the latitude, and ω the angular velocity of the spin of the earth (ie 15° per hour; or 7.29×10^{-5} radians per second):

$$F = 2\omega.V \sin \theta$$

Fig. 1.7 explains the latitudinal variation. At P_1, at latitude θ, the direction in space of true North is the tangent P_1N. After the earth has made half a revolution this point is at P_2, and the direction of true North in space has changed to P_2N. The angle between these directions is 2θ. At the equator (latitude 0°) F becomes zero.

This so-called Coriolis force is a mathematical convenience to show the apparent deflection due to the earth's rotation. Fig. 1.7 shows that, in consequence, the air may eventually flow *along*, rather than as a gradient wind *across* the parallel isobars. It has become a **geostrophic wind**, its speed proportional to the spacing of the isobars.

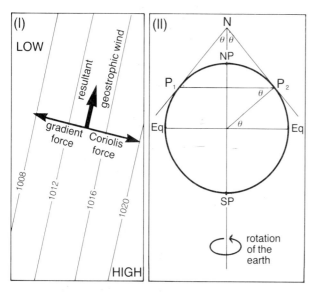

Fig. 1.7 The Coriolis force and the geostrophic wind.

1.5 Surface frictional drag

Near the earth's rough surface the frictional drag effectively acts against the Coriolis force. But high in the atmosphere, where it is absent, air tends to flow parallel to the isobars. The result is a wind spiral with height–**the Ekman spiral** (Fig. 1.8).

1.6 Air flows about centres of low and high pressure

The Coriolis force tends to cause air to move parallel to the isobars, so the winds encircle low pressure centres. But the **frictional force** acts

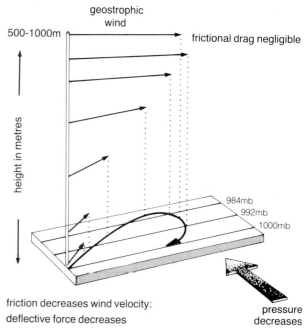

Fig. 1.8 The Ekman spiral.

against the air flow, reducing its speed. As the Coriolis force is proportional to wind speed, it, too, is reduced. This tends to make the air flow across the isobars, towards the centre of low pressure.

Over a smooth ocean surface the deflection is usually less than 10°, though a rough land surface can cause a deflection of up to 30° or so towards the centre. Air movements from a high pressure area are generally lighter, but are subject to similar frictional modification.

Centrifugal force also acts to increase the speed of air flowing from a centre of high pressure, and acts to oppose and decrease the speed of air as it moves about a centre of low pressure. But this force only becomes significant for intense low pressure systems when there are winds of high velocity.

Fig. 1.9 Conflicting forces affect air circulation.

CIRCULATION ABOUT A LOW PRESSURE SYSTEM

P gradient pressure
G geostrophic wind
F frictional force
C Coriolis force (reduced by friction)

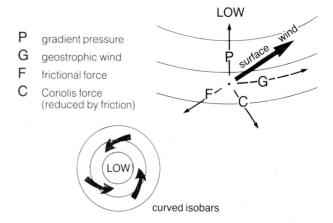

2

ENERGY EXCHANGES:
EARTH AND ATMOSPHERE

2.1   Energy exchanges

The **solar energy** arriving at the outer limits of the earth–atmosphere system interacts with the various components of the atmosphere as it passes through it, and then with the very varied features of the earth's surface. Some is reflected back to space; some is absorbed and transformed into heat, and so is again emitted as long-wave radiation. In the case of water, the energy absorbed may cause a change of state from solid to liquid to vapour. In other words some is stored, in the short term, as potential energy in vapour, and released on condensation to droplets. Some brings about biochemical processes in plants and so is temporarily stored in vegetation, and perhaps transferred to animals. Some is stored for long periods in fossil fuels, in coal and oil, and may be released into the atmosphere by human activities.

2.2   The global energy balance

The exchanges by which the earth–atmosphere energy balance is maintained on an annual scale are summarised in Fig. 2.1. The figures are, of course, only approximations of the average conditions for both earth and atmosphere. They omit the effects of vegetation, topographical differences, and daily and seasonal variations. And they are not concerned with the large-scale exchanges of energy *within* the system.

As we have already seen, as solar energy passes through the atmosphere some is absorbed by gases. Some is scattered by the molecules and reflected back to space; some is scattered but continues to travel within the system. That scattered by atmospheric molecules is mostly at the blue end of the spectrum, which causes the sky to appear blue by day.

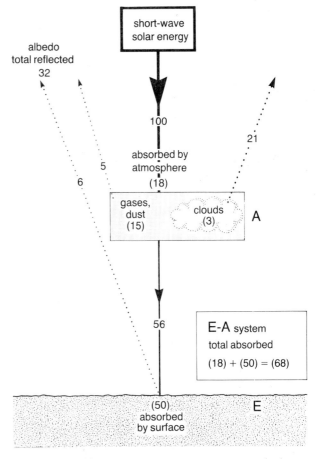

Fig. 2.1   The earth–atmosphere system absorbs about 68 per cent of the insolation received at the outer atmosphere. The remainder is reflected back as the total E–A albedo.

Eventually about a third is reflected back from clouds and from the surface. The proportion lost by scattering and reflection is the average **global albedo**. Of course the percentage of short-wave

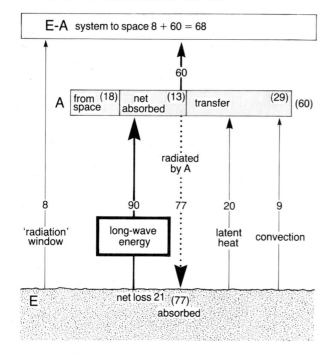

Fig. 2.2   The energy absorbed (Fig. 2.1) is returned to space partly through the 'radiation windows', shown in Fig. 1.3, but mainly from the atmosphere (after transfers of heat energy between earth and atmosphere).

radiation reflected varies locally, for different surfaces have their own albedo (p.142).

As the surface absorbs energy its temperature rises. Like any other hot body, it also radiates energy. As this is of long wavelength, it is readily absorbed by the lower atmosphere, especially by water vapour and water droplets (clouds), and by carbon dioxide. These in turn are heated and emit radiation in all directions. Much is re-radiated back to the surface; and, of course, clouds also effectively close the radiation windows (p.3). The result is what is termed a **'greenhouse effect'**, which acts to maintain fairly high air temperatures near the surface ... though a greenhouse also prevents energy losses through air turbulence.

The overall picture, then, is that the troposphere is relatively transparent to short-wave radiation and gains energy mainly from long-wave radiation from the surface. The importance of water vapour and cloud cover must be stressed. About a fifth of the short-wave reflection, the global albedo, is from clouds; they absorb much infra-red radiation which would be lost to space and they also re-radiate energy to the lower atmosphere.

The fact that there is an overall balance between energy received and lost to space does not mean that there are uniform conditions in the troposphere. The differences in radiant energy gained and lost by different parts of the earth cause energy to be transferred from one part to another, and create large-scale and small-scale atmospheric

circulations (Chapter 4). We should, therefore, consider the surface properties and seasonal variations responsible for the continuous interrelated air flows over the earth's surface.

2.3   Latitudinal differences

The **solar elevation** varies with latitude and seasons. Rays falling obliquely to the surface give less energy per unit area than those striking vertically, for the energy is spread over a larger area. Also, as Fig. 2.3 shows, the length of the passage through the lower atmosphere varies with latitude. A longer passage means increased absorption, scattering, and reflection.

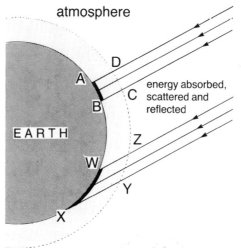

WX > AB :   energy received per unit area of the surface is greater at AB than at WX

ZW > DA :   more energy lost during longer passage through the atmosphere

Fig. 2.3   The energy received per unit area varies with latitude.

The energy received varies with the length of daylight period, and thus with the latitude and seasons. At the equator the longest period of continuous daylight is 12 hours; but the elevation of the noonday sun is high throughout the year. Within the Arctic and Antarctic circles the longest daylight period varies from 24 hrs at latitude 66½° to six months at the pole; but the noonday sun is never within 40° of the zenith, and is at a low elevation for much of the year. Nevertheless, as Fig. 2.4 shows, during mid-summer the continuous insolation produces a daily input in excess of that received on the equator at that time. During the long period of darkness the loss by radiation continues, so that, all else being equal, temperatures fall progressively during the winter months.

HEAT ENERGY PER UNIT AREA FROM INSOLATION

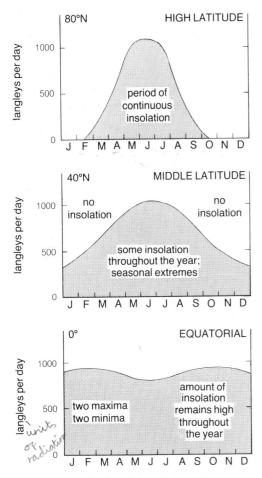

Fig. 2.4 Latitudinal variations in insolation.

Elsewhere over the globe there are various combinations of shorter winter days, with the sun relatively low in the sky, and longer summer ones, with a higher noonday sun. Fig. 2.4 shows the combined effects of length of day and elevation of the noonday sun on the energy received at the outer limits of the atmosphere in certain latitudes.

2.4 Surface conditions and energy transfers

The **specific heat** of land surfaces is less than that of water. This means that for a given amount of energy received their temperature rises more rapidly than that of an equivalent mass of water; and as they lose heat by radiation, their temperature falls more rapidly.

In water **convection** currents can transport energy freely through the body of the liquid. Sometimes conditions may not favour vertical mixing to any great depth, for heated, less dense, surface water may overlie cooler, denser water. Nevertheless, currents, set in motion by wind energy, transfer top-water horizontally and may allow lower water to well up. By contrast, the amount of heat transferred by **conduction** from a solid surface through the ground is very small.

Land surfaces heat and cool quickly. In hot deserts the surface temperature may be of the order of 80°C; by night temperatures fall sharply. The resulting physical stresses, and the tendency for dew to form on the surface at night (p.16), are potent factors in rock weathering. By day the air a metre or so up may be 30 C° cooler than that in contact with the surface; the density differences can create swirling 'dust devils' and winds bearing particles capable of eroding exposed rocks.

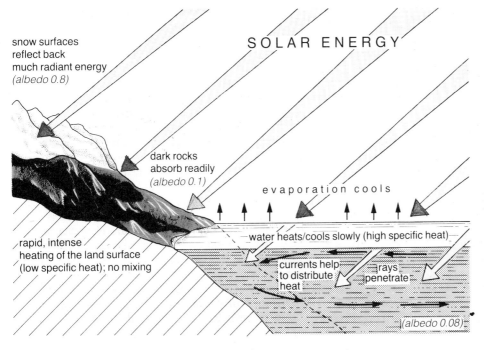

Fig. 2.5 The differences in the absorption and distribution of radiant energy affect the surface temperatures and those of the air in contact. The albedo of water increases as the angle of the sun's rays decreases from the zenith. The albedo is sometimes expressed as the percentage reflected back (ie snow, 80 per cent).

Because water distributes energy through its mass, it not only has a slower rise of surface temperature, but cools more slowly, and as a whole: for as cooling water increases in density (down to 4°C) it sinks and is replaced by warmer water from beneath. In a pond or lake this may continue until the whole mass is gradually cooled. Oceans are sometimes referred to as **heat reservoirs**, for their supply of heat is available to the atmosphere for a longer period than that from land surfaces. Land does not store heat in this way, and does not have such long-term effects on the atmosphere. This means that there are often striking differences between the air temperatures over continents and those over oceans.

Evaporation from open water and moist land cools the evaporating surfaces; the extent to which it does so depends on the existing temperature and air humidity. Dry land, obviously, cannot share this effect.

The effects of transparency and surface reflection (**albedo**) also vary with the nature of the surface. Water in the liquid state allows energy to penetrate to great depths. Water as ice and snow, with its high albedo, reflects back some four-fifths of the solar energy received, and warms slowly; whereas a grassy land surface may reflect only a fifth, and a dry black soil reflect as little as a tenth, and warm more rapidly.

2.5   Land breezes and sea breezes

The differential heating and cooling of air over land and water surfaces can give rise to land breezes and sea breezes. Near seas, or large lakes, when the weather is warm or hot, and conditions calm, an onshore wind sets in during the day, but is replaced by an offshore breeze at night. Generally the onshore air-flow is stronger than the offshore one.

Fig. 2.6   Air circulations under hot, relatively calm conditions.

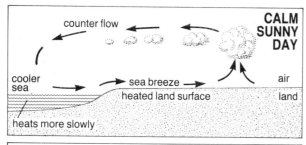

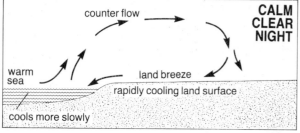

Fig. 2.7   Cumulus clouds develop as hot air rises in convectional updraughts (thermals) over the northern coastland of Tasmania. The sky remains clear over the cooler waters of Bass Strait.

By day the near-coastal land surfaces become much hotter than the adjacent water surface. Air heated by the land expands and rises. A pressure gradient is established between the cooler, denser air over the water and that over the land, thus causing an onshore sea breeze. The flow is reversed aloft, with air moving towards the water. Clouds formed in the bubbles of rising air may thus drift slowly towards the coast, gradually dispersing as they move.

In the tropics strong sea breezes may move as far as 100 km inland. They may bring low-level sea-mist (which evaporates and clears over the hotter inland surface) or form cloud, or even precipitation, as the air rises on heating. In temperate latitudes the effects of sea breezes are less noticeable and they are often overshadowed by general light winds.

On clear, still nights, land temperatures fall rapidly to below those of the water surface. Air cooled from the land surface drifts towards the water as a land breeze.

2.6   Topographical influences on overlying air

The surface temperature of undulating ground varies as the sun's rays strike some places at oblique angles, and at other places vertically. Sunlit slopes may be remarkably hotter than shaded ones.

In most latitudes winds can occur in mountain valleys under otherwise still conditions. In north-south valleys in the northern hemisphere, on still

Fig. 2.8 Inversion conditions as air subsides over the western central Andes. Cloud forms in cold, dense air in the deep valley. Further east, high cloud is forming where moist, unstable air from the Amazon basin is rising up the steep eastern slopes.

sunny days the air above the valley slopes, heated by the sunlit surface, may become much warmer than the air overlying the centre of the valley, and move up-slope as shallow **anabatic** draughts. Air over the centre of the valley sinks and maintains the circulation. This may be accompanied by an **up-valley wind**, created by cooler air moving in from an adjacent plain. This valley wind is generally overlain by an upper return current.

The south-facing slope of an east-west valley in the northern hemisphere receives more radiation than the north-facing one, and may experience strong up-slope draughts on hot summer days. The shaded slopes tend to be moister, with a lower tree-line, and in winter retain snow for longer periods.

On cold, still nights the air is chilled from surfaces continuously losing heat by radiation, especially from the upper slopes. The cold dense air flows as a **katabatic** wind down the slopes and collects above the valley floor, often creating a frost hazard for vegetation within the valley or on the lower slopes. As cold air lies close to the ground with warmer air above, the normal fall in temperature with altitude is reversed, a condition known as **temperature inversion**, here on a local scale. In an open valley the cold air may continue to flow down-valley onto adjacent lowland, as a **mountain wind**, with a return current higher up above the valley.

In low latitudes the floor of broad valleys can become excessively hot by day, causing very powerful up-draughts, hazardous for small aircraft.

Where, by contrast, surfaces are much colder than their surroundings, as on an elevated ice-cap, cold, dense air may accumulate above and drain to lower areas by gravitational pull. In Antarctica and Greenland this downward flow sometimes

Fig. 2.9 Anabatic and katabatic air flows, with induced mountain and valley winds.

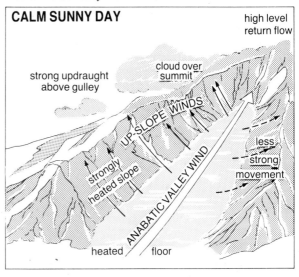

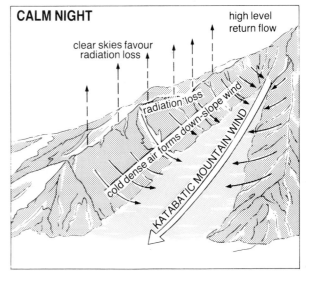

becomes a really strong wind (a **gravity wind**). Even in mountain country with small isolated ice surfaces, cold winds of this type are noticeable. In fact, mountain climates are extremely variable; high surfaces, where the atmosphere is thin, are strongly heated and lose long-wave radiation very rapidly; while valleys may experience inversion conditions during clear calm weather, but a greenhouse effect when blanketed by low cloud.

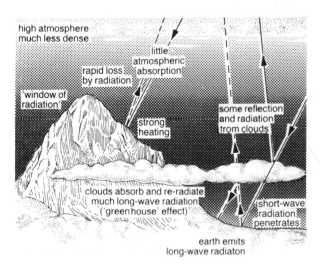

Fig. 2.10 High mountains rise to altitudes where the atmosphere is thin and there are neither the persistent clouds nor dust particles which occur at lower levels. Consequently they act as windows of radiation.

2.7 Natural obstacles to air flow

Of course every hill and depression will disturb the pattern of air flow. A flow of air becomes a high velocity stream as it encounters a valley constriction or a gap in a hedge, increasing its speed to take the mass of air through the narrows.

When air flows over and round an isolated hill, it must similarly accelerate over the upwind side. But it is retarded by friction, so that as it spreads again and slows beyond the hill it leaves a zone of relatively low pressure in the lee, where **wind eddies** develop. These are apt to bring air pollutants from hillside sources back to ground level.

A similar low pressure zone in the lee of a long mountain range tends to create waves in the atmosphere, with clouds forming at the crests as they extend away from the lee slopes in long parallel lines (p.22).

In fact **friction eddies** occur when strong winds pass over any rough surface. The drag of air over the irregularities creates **surface shearing stress**. This sharply decreases the horizontal wind speed. The effect lessens as it extends upward towards the boundary layer (p.159), above which air flows at its normal speed. Eddies generated by the friction also check the flow below the boundary

Fig. 2.11 Moist air rising up the conical slopes of Mt Egmont in the North Island of New Zealand creates a persistent saucer-like layer cloud. Under stable conditions the descending air warms, and the skies around the massive volcano are clear.

layer. The whole pattern may be disturbed, however, when heat-generated currents make the lower air turbulent, especially when the air is unstable (p.18).

Fig. 2.12 The velocity of a mass of air flowing with a given momentum increases as it becomes restricted by a narrow channel (A). An obstacle disturbing the flow creates increases and decreases of pressure, causing eddies to form as the air readjusts, as shown in profile B and plan C.

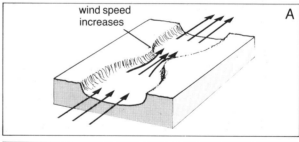

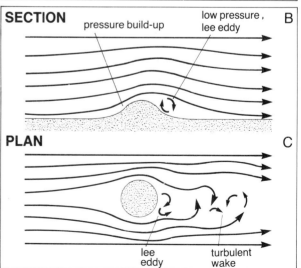

Strong turbulent wind can dry the surface of agricultural land, erode the top-soil, and transport particles long distances. Where the wind blows from one main direction, a **shelter belt** of trees may be established as a barrier.

DENSE BARRIER A

fast air drawn in

h

influence 10-15h downwind

eddies

LESS DENSE BARRIER B

cushioning

h

through flow

influence 20-30h downwind

h   height of shelter belt

Fig. 2.13 A dense windbreak of height h reduces the wind speed but creates eddying: air drawn in causes the velocity to be regained at a distance of about *10h* downstream. When there is a low through-flow (B), the velocity is partly checked, and without eddying the influence of the wind-break extends further down-wind.

As in Fig. 2.13, large-scale eddying will occur in the lee of a solid barrier, with lesser eddies tailing-off downwind. These may also adversely affect the top-soil and crops. However, a reduction of wind speed of at least a tenth of its value occurs downwind to about 10–15 times the height of the obstacle.

A tree belt which allows a moderate through-flow of air at lower trunk level, but provides an effective barrier at crown level, prevents large-scale eddying. It still reduces the wind speed and extends the protected zone further downwind, perhaps to 20–30 times the height of the tree-belt.

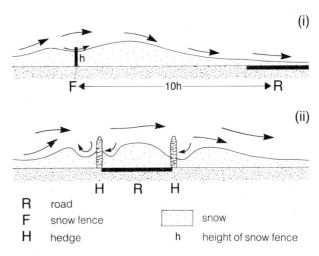

(i)

h

F ◄——————— 10h ———————► R

(ii)

H   R   H

R	road		
F	snow fence	▢	snow
H	hedge	h	height of snow fence

2.14 A low fence placed some distance up-wind parallel to a road creates local drifts and affords some protection to the road. Close parallel hedges may not afford protection, and may, in fact, cause overlapping drifts.

Extensive shelter belts of trees tend to increase precipitation, allow higher dewfall in the more sheltered areas, and reduce the water loss through evapotranspiration. They also help to retain a snow cover; and in the temperate continental interiors the ability of snow to insulate soil from severe frost, and the ability to release a large quantity of water on melting, are advantages.

Snow-fences set back from a road can cause snow to accumulate in the lee of the fence, and allow the wind to sweep the area beyond relatively clear of snow. But, as Fig. 2.14 shows, parallel hedges bordering a road can lead to accumulation on the road surface and create snow blocks.

3

MOISTURE AND
ATMOSPHERIC STABILITY

3.1   The cycling of atmospheric moisture

The moisture content of the air has a great influence on the nature of air movements in the troposphere.

At any moment over 97 per cent of the planet's water is held in the oceans, though, like all water in the E–A system, it is part of continuous circulation – **the hydrological cycle** – fuelled by the input of energy from the sun.

Evaporation transfers liquid water from the surface into the atmosphere, as water vapour; a process which is also achieved by **sublimation** (direct from solid to vapour) from ice, and **transpiration** from plants. The water vapour may condense to become liquid, and remain suspended in the air as droplets – cloud, fog, or mist; or it may be precipitated, falling back to the surface as rain,

Fig. 3.1   The hydrological cycle: showing the sources of energy involved in transferring water as solid, liquid, or vapour from one part of the world to another, and indicating the percentage held at any moment in each state, in various locations.

hail, or snow, or as dew on cold surfaces – all forms of precipitation.

Some water will be flowing in streams, rivers, and floods, finding its way back to the ocean as surface run-off. But during this circulation other water is present for various lengths of time beneath the ground in soil and rocks, in plants and animals, and, temporarily immobile, in the body of lakes, oceans and masses of ice (**stored water**).

Figs 3.1 and 3.2 show the hydrological cycle, with the storage, and the percentage actually being exchanged between the surface and atmosphere. The proportion of water vapour seems small, but its equivalent as rain would cover the entire earth to a depth of 25.5 mm.

As with heat, there is a water balance, with roughly a constant amount of water vapour in the atmosphere. But, of course, there are great inequalities of precipitation and evaporation over various parts of the land surface and oceans.

As we have seen, water in the atmosphere is the chief absorber of incoming radiation and, especially, of radiation emitted from the surface. Dry air with clear skies allows more energy to be received

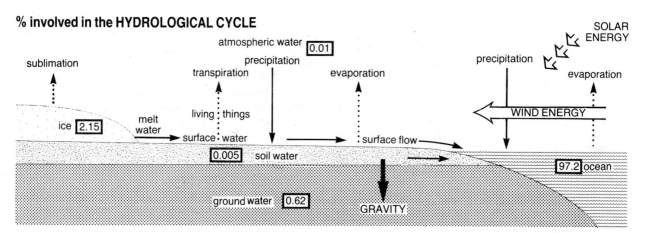

% involved in the HYDROLOGICAL CYCLE

% FLOWING IN CYCLE (-)

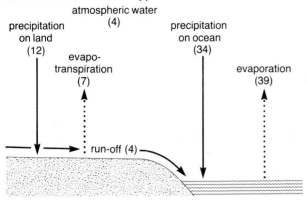

Fig. 3.2 Water being transferred between exposed surfaces and the atmosphere.

at the surface and much long-wave radiation to be lost. This makes for a large temperature range in the lower atmosphere. When there is low cloud cover the droplets reflect and re-radiate energy towards the surface and reduce direct surface heating. In Britain, when there is 8/8 low cloud cover the daily variation in air temperature may be only 1–2 C°; but with clear skies the range can be ten or fifteen times as great.

3.2 Absolute humidity and relative humidity

The amount of water vapour present in the air at any place varies greatly. The water vapour content of the lower air tends to be some 2 per cent of its volume; but varies from very little to about 5 per cent. The amount present also tends to decrease upward, so that the mean vapour content at 1200 m is only about a tenth of that near sea level; though, again, the actual amount at any height is always varying.

The actual mass of water vapour in a unit volume of air (g/m³) is the **absolute humidity**; though, rather than use volume measurements, meteorologists usually refer to the mass of water vapour in a given mass of dry air (g/kg) as the **mass mixing ratio** (x).

The water vapour content may also be stated as the proportion of the atmospheric pressure due to the water vapour content, say 20 mb out of a total atmospheric pressure of 1000 mb. Cold, dry air might have a water vapour pressure of less than 2 mb; whereas in warm, moist, tropical air water vapour might exert a pressure of 15–20 mb.

At any given air temperature there is a limit to the amount of water that can be held as vapour. Once this limit of **saturation** is exceeded, **condensation** usually occurs.

The absolute humidity of the air does not change unless water is added to, or taken from, the body of air concerned; but the relative humidity varies with the temperature. **Relative humidity** is defined as the proportion of the actual mass of

water vapour contained in a given volume of air to the maximum amount that could be contained *at that temperature*. It can be expressed as a percentage.

As the water vapour may be expressed in terms of the pressure it exerts, a comparison of the actual vapour pressure with that which would be exerted if the air were saturated at that temperature also gives us the relative humidity.

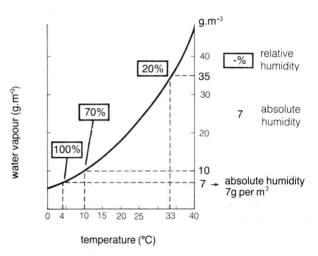

Fig. 3.3 Relative humidity: variations with temperature.

Consider air which is saturated at 4°C: its relative humidity is 100 per cent. As it warms the value falls to 70 per cent at 10°C; to 50 per cent at 18°C; and as low as 20 per cent at 33°C (Fig. 3.3). So very dry air over a hot desert may contain as much, or more, water vapour as saturated air over Arctic waters.

3.3 The process of condensation

Unsaturated air may become saturated by decreasing the temperature; or, if the temperature remains constant, by increasing the water vapour content.

When unsaturated air is cooled, the relative humidity increases until it is completely saturated. This is the **dew-point**. Any further cooling leads to condensation. If the dew-point is below 0°C, some of the condensation may appear as ice crystals, as snow, white frost, or high (cirrus) clouds.

In practice liquid condensation often occurs at temperatures well below freezing point. In order for water droplets to form in the atmosphere, it is necessary for tiny particles to act as **nuclei of condensation**. These may be microscopically small, such as dust particles or salt crystals, which are common in the lower troposphere. Particles with an affinity for water (hygroscopic substances) are the most effective nuclei of condensation. Salt particles over the ocean are particularly suitable, and, of course, plentiful.

Where nuclei are absent, cooling may continue far below the dew-point without condensation, and the air becomes **super-saturated** with water vapour. Such conditions are obviously less likely to occur below, say, 6000 m than in the higher, purer atmosphere.

3.4   Changes of form and state

Water frequently changes its state, from solid to liquid to vapour. In order to do this it must receive energy. The amount required per unit mass for melting (at constant temperature and pressure) is **the latent heat of fusion**, which is 334 joules (80 calories) per gramme of ice at 0°C. That required for evaporation (**the latent heat of evaporation**) is 2257 joules (540 calories) per gramme.

When water vapour condenses to droplets, and liquid water freezes, equivalent amounts of latent heat are released to the surroundings.

3.5   The nature of precipitation

The term 'precipitation' includes such condensation forms as rain, snow, sleet, hail, dew, hoarfrost and rime.

When air is caused to rise it expands and cools. At a certain height it becomes saturated with water vapour and condensation occurs. But condensation releases heat energy, which counteracts the cooling by expansion. This may keep the air warmer and more buoyant than the surrounding air, so it continues to rise. Condensation may occur on a scale sufficient to bring precipitation to ground level.

Droplets are formed, usually of radius between 0.001–0.05 mm. Very small droplets tend to avoid one another if set in motion. But slightly larger ones may collide and coalesce into larger and larger drops. Eventually they are able to fall to earth as **rain**, often against considerable updraughts. The maximum size of a raindrop is of the order of 5 mm radius.

The rate of fall of droplets varies. A drop of 0.5 mm may take four minutes to fall one kilometre; one of 2.0 mm radius may take less than two minutes. Very fine rain may evaporate below the cloud base before reaching the ground.

The term 'warm rain' implies that it falls from clouds with no ice crystals, 'cold rain' comes from ice crystals which have melted.

When temperatures fall below freezing point, ice crystals may be formed as droplets freeze; though sometimes crystals form directly from vapour. As the crystals coalesce they fall as **snow**. Very cold air has little moisture, so heavy falls are unlikely. But considerable falls can occur when warm moist air meets a mass of very cold air. A warm airmass encroaching on cold anticyclonic air in mid-winter (as a warm front, p.37) usually gives a very heavy fall, with large flakes in the warmer air. Fine, hard snow is more typical of cold conditions. The approximate equivalent of 300 mm of light, packed snow is 25 mm of rain.

Hail is usually formed in towering cumulonimbus clouds (p.20). Strong updraughts carry droplets to high altitudes, where they freeze within the cloud. As they fall, further condensation takes place on the particles before they are again carried up by air currents. The process may be repeated many times before the resulting hailstones escape the main updraught and become large enough to fall to earth. When they do fall they are cushioned to some extent by the uprush of air. The hailstone's interior shows concentric shells of ice.

Dense masses of droplets formed by condensation and held in air close to the ground comprise **mist** or **fog**. (When the droplets restrict visibility to less than a kilometre, this is regarded as fog rather than mist.) They form when moist air is cooled below its dew-point, or sometimes when extra water vapour is added to a mass of air from a surface water source – a lake or river, or moist vegetation.

When clear skies freely allow radiation loss and the air in contact with the ground is chilled to below dew-point, condensation occurs as **radiation fog**. Gravity plays a part in building up such fog belts, as when valleys receive cold, dense air from higher ground. The fog develops and thickens from the bottom upward. Here, again, there is local inversion, with cold air close to the ground and warmer air above.

Sometimes meteorological conditions cause air to subside from high altitudes and warm by compression. This gives rise to more general temperature inversion, and any fog formed in the colder air beneath may persist for long periods.

At other times fog may lift and form low cloud layers during the day, before lowering and thick-

Fig. 3.4 Humidity increases as a depression approaches western Cyprus: condensation causes clouds to form high on Buffavento in the northern mountains; beyond, the plains still swelter in the sun.

ening again at night. Much of what is known as 'hill fog' is in fact cloud at hill level.

Masses of air moving horizontally above the surface, as **advection currents**, transfer heat energy and moisture, so that fog may form where warm, moist air moves over a cold surface. Dense **advection fog** frequently forms off the Newfoundland coast, where warm air from above the Gulf Stream passes over the cool waters of the Labrador Current. While such fogs are common during winter along sea coasts, they also occur when warm air moves in over very cold surfaces, especially snow-covered land. Advection fog is common in middle latitudes when warm air displaces a winter anticyclone (p.36) which has given a cold spell of weather.

Hoar frost occurs when the temperature is well below freezing and the water vapour is super-cooled. Minute ice crystals are deposited on grass, leaves and cold surfaces, though there may also be frozen drops of super-cooled water. **Rime** is usually a heavier deposit (ice crystals which build up on the windward side of objects), formed when the air is moving very slightly, often when foggy conditions occur under clear skies.

Clouds, too, are masses of tiny suspended droplets. But the form of clouds varies a great deal and reflects conditions in the lower atmosphere. So, before considering cloud forms in detail (p.19), it is necessary to appreciate how the properties of air change from the surface upward through the troposphere – the vertical distribution of temperature and humidity.

3.6   Humidity and the temperature lapse rate

The rate at which air temperature decreases with altitude is known as the **lapse rate**.

As the chief source of heat is the earth's surface, the lapse rate in air close to the ground often far exceeds the normal rate. The first hundred millimetres or so tends to have a micro-climate of its own (p.103). A temperature fall of 20 C° in the first metre above a hot desert surface is usual, compared with the average figure for the normal lapse rate of 0.6 C° per 100 metres. On a sunny afternoon in the tropics a steep gradient may exist up to several hundred metres above the surface.

The lapse rate varies from place to place, with the season, the time of day, and the water vapour content, and there is almost always a variation in lapse rate with altitude. The actual temperature decrease with height in the air about you, or in the air surrounding a particular mass of air under observation, is the **environmental lapse rate** (ELR).

Now consider an unsaturated mass of air in contact with the surface and what happens when, for some reason, it is displaced and rises through the surrounding atmosphere. We may assume that no heat is transferred between this air and the environmental air, a circumstance known as **adiabatic**.

The atmosphere is never completely dry; but unsaturated air is described as 'dry' if its moisture content does not condense, and so affect the lapse rate. Its cooling rate, the **dry adiabatic lapse rate** (DALR) is about 1.0 C° per 100 metres.

As this air rises and is chilled to saturation point (its dew-point), much of the water vapour condenses to form droplets. This liberates latent heat energy. Heavy condensation thus has a considerable heating effect, which acts against the adiabatic cooling. The rising, now saturated, air thus cools at a slower rate – the **saturated adiabatic lapse rate** (SALR).

At first the SALR is only about half the rate in dry air, about 0.5 C° per 100 metres. But this is a variable rate. Higher up, with lower temperatures and a low moisture content, the rate increases. At temperatures between 0° C and −40° C the rate is about 0.75 C° per 100 metres. At about 12 000 m the value may be close to the DALR.

We have already considered **inversion** conditions, where warmer air overlies cooler and produces a negative lapse rate. This occurs on a large

Fig. 3.5 High on Exmoor, during a bitterly cold spell, damp air forms rime on the trees, and at noon icicles still hang from the gate.

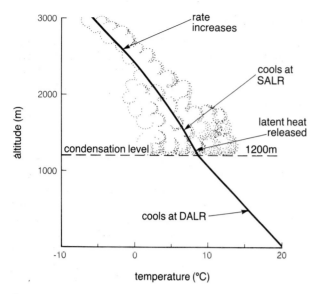

Fig. 3.6   Moist air is forced to rise and cool; but the rate of cooling is checked by condensation and energy release. The shape of the SALR curve shows that the cooling rate tends to increase again with altitude.

scale in the relatively stationary high pressure systems in the sub-tropics, above the great deserts, and over the continental interiors in winter. But less permanent inversions occur occasionally in changing weather systems. From time to time masses of cold air move beneath a layer of warmer air, which persists aloft, as an occlusion (p.37).

Advance knowledge of the lapse rates within advancing air masses is essential to forecasting. They may indicate the likelihood of strong vertical movements, with possible condensation, cloud formation, and rain; or may point to more stable conditions.

Fig. 3.7   Relative stability affects the behaviour of air forced to rise, or remain close to the surface as in D.

3.7   Stable and unstable air

Vertical air movements are involved in changing weather conditions. Consider again a parcel of air which is caused to move vertically. Should the differences between its properties and those of the surrounding air result in its further upward movement, the local atmosphere is in an **unstable** condition. Should it be forced to rise vertically, but the vertical movement is so resisted that it tends to return to its former level, then the **local atmosphere is stable.**

Fig. 3.7A shows air forced to rise, but cooling at the dry adiabatic rate. It remains cooler, and so denser than the surrounding air. After displacement it sinks to a lower level: a state of **stability**.

Fig. 3.7B shows rising air which has a lower lapse rate than the surrounding air. So long as it remains warmer and less dense than the environmental air, it continues to surge buoyantly upward. This is **absolute instability**, which is, in fact, a rare state.

Fig. 3.7C shows air which is forced to rise, perhaps by encountering a mountain side or a mass of cold dense air. It begins to rise and cool at the DALR, and remains cooler than the surrounding air. But at a certain height condensation occurs. This releases heat, with the result that the air cools less rapidly. Eventually it becomes warmer than the environment, and hence unstable. It now continues to rise by its own buoyancy. Such **conditional instability** occurs much more frequently than absolute instability. In fact air of only moderate potential instability may not begin to rise until it encounters an obstacle.

In general, air approaching saturation, when it would be subject to a lower rate of cooling, is likely to be less stable than unsaturated air. Also, air containing water vapour is *lighter* than the same volume of dry air (water vapour being lighter

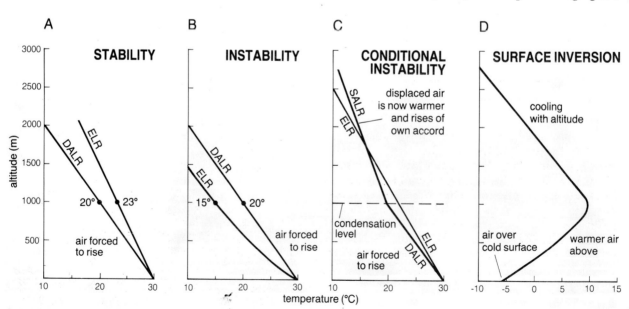

than the atmospheric gases). Such warm, damp air, rising to great heights may well create the towering clouds we see on a hot summer's day, and so frequently in humid conditions within the tropics.

Fig. 3.8 shows part of a **tephigram**, which can be used to determine the condensation level. This would normally also include lines related to the saturated mixing ratio. Here it indicates a state of conditional instability. The heavy wavering line shows the environmental lapse rate as observed and recorded at a particular time. The line AC indicates air ascending from A and cooling at the DALR. At C, the condensation level, its temperature is below that of the environmental air. CB shows the now-saturated mass of displaced air cooling at the SALR.

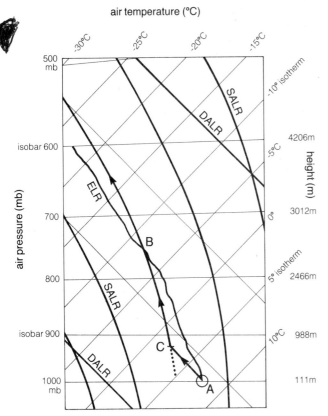

Fig. 3.8 A tephigram showing conditional instability.

At B its temperature is that of the environmental air. Up to this point there has been a stable situation. But above B the state of the local atmosphere is such that the rising mass of saturated air remains warmer and more buoyant, and so surges upward. It may rise to great heights and produce towering clouds. The latter, as opposed to layer clouds, suggest some degree of instability. Clouds generally are useful indicators of relative stability.

3.8 Cloud formation

Clouds consist of countless millions of very tiny droplets or minute ice crystals, maintained in the atmosphere by slight upward movements. When the sun shines on their outer surfaces, or through thin layers, they appear white by reflected or refracted light. The sides appear grey, or black when the cloud is dense.

Cloud forms tell us a great deal about atmospheric conditions, and we may be able to estimate whether clouds in the lower atmosphere are building or dispersing by observing their outer edges. Sometimes clouds form in an up-current and disperse in a down-current, as turbulent conditions with eddying may cause humid air to rise and fall above and below condensation level. In other words, behaviour as well as shape can also reveal a great deal about the atmosphere.

Low Clouds (from the surface to about 2000 m)

Stratus is a low, dense, uniform, dark grey layer of droplets, similar to blanketing ground-level fog. It does not rest on the ground, and often forms well above the surface. Ragged in appearance, and usually shallow, it may drift swiftly along. Stratus may form locally in moisture-laden air, even when

Fig. 3.9 Types of cloud.

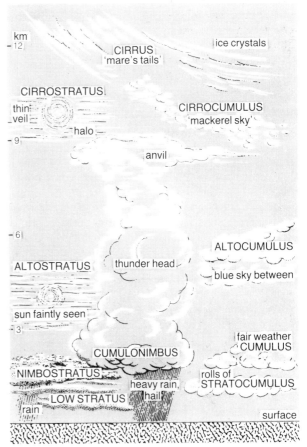

Fig. 3.10 Convection bubbles rise from the central Australian tableland, north of the Macdonnell ranges, and form billowing masses of cumulus cloud.

Fig. 3.11 A ground view of joined cumulus cloud formation above the condensation level, giving scattered showers over the Australian bush, north of the Macdonnell ranges.

weather conditions are otherwise fair, blotting out a hillside or hanging close to a summit. Thick stratus associated with rain or snow is known as **nimbo-stratus**.

Stratocumulus clouds are usually low, soft-looking masses, with a somewhat globular or roll-shaped appearance. Open sky shows between them. There is often a regular pattern, particularly when rolls of cloud form at right angles to the general direction of cloud movement. This type of cloud is mostly associated with clearing weather.

Clouds with marked vertical development (rising to higher levels)

Cumulus of the fair weather type is formed when convection is strong (p.19). Bubbles of warm air (**thermals**) rise from the heated surface. In some the water vapour condenses at a particular height, indicated by the flat base of the small billowing cloud. A few may grow in size and develop vertically until more stable conditions occur at higher levels. Individual thermals form separate updraughts within the cloud, so that the billowing, domed masses of the upper surface resemble a cauliflower. Where illuminated by the sun, they appear white, but grey on the shaded sides. As convection currents lose their strength, usually towards evening, the clouds tend to die away.

Cumulonimbus. In these the air remains unstable to considerable heights. The vertical development is great. Clouds may extend from about 500 m to towering summits at over 10 000 m, their tops dazzlingly white. The thick shaded masses give the threatening appearance of a thunder-cloud seen from the ground.

Super-cooled droplets and ice crystals form in the higher parts of these clouds, releasing more latent heat. This helps to increase the rate of upward movement. Strong, cold downdraughts about the cloud contrast with the warmer air within and also accelerate the updraughts. Eventually a layer of ice crystals (cirrostratus) forms at the top, spreading out in the typical anvil shape as it encounters the strong inversion of the tropopause. Winds carry the ice particles forward.

Torrential rain and hail often result from these movements. The downdraughts reach the cloud base and spread out in cold gusts, often against the surface wind (Fig. 3.13).

Lightning and thunder frequently occur, with discharges of accumulated static electricity between cloud and cloud, or cloud and earth. The discharge heats the air and causes great expansion. The resultant contraction produces the initial sound of thunder.

Fig. 3.12 Condensation occurs in the individual bubbling updraughts of warm, moist air, developing into 'cauliflower' masses of cumulus.

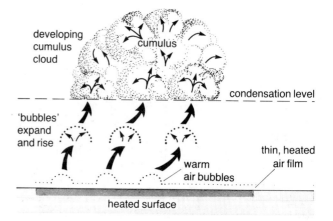

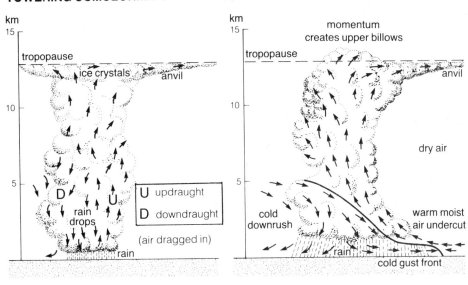

TOWERING CUMULONIMBUS

STORM WITH COLD GUST

Fig. 3.13 The downflow of air, which counters the strong updraughts, drags in surrounding air. This can create a strong surface wind which lashes rain forward against air moving into the system

The charges result from the updraught. These tend to separate positive and negative charges, possibly helped by the splitting of an ice shell about water droplets. The charges become concentrated in certain parts of the cloud.

Cumulonimbus clouds do not only build up in unstable air on hot sunny days. Sometimes a stream of cold air, or the base of a polar airmass (p.41), receives energy and moisture from a warm sea. There is then a high lapse rate and vigorous convection. If a warm land surface, or outstanding relief, then provides additional uplift, very heavy rain may occur.

Middle clouds (usually formed between 2000 and 6000 m)

Altostratus is a thick cloud layer, which may merge with higher cirrostratus; but at this lower level the more abundant water vapour gives a dense dark cloud, through which the sun may faintly gleam. It forms most frequently where uplift and condensation occur in advance of the warm front of a depression (p.37), and usually blankets the sky. Fairly steady precipitation follows its development.

Altocumulus generally forms under fair weather conditions, and the cloud masses have some vertical development. Blue sky appears between lines or layers of individual clouds. These can be caused by layers of air of different density and humidity flowing over one another, producing a billowing effect. The clouds appear white, or grey on the shaded sides.

High clouds (forming generally above 6000 m)

Cirrus clouds are composed of ice crystals and are usually wispy or feather-like in appearance. They

Fig. 3.14a Vigorous updraughts which rise from islands in a Swedish lake flatten out on reaching the layer of strato-cumulus cloud formed under inversion conditions.

Fig. 3.14b Some of the rising air penetrates the stable layer and forms billowing cumulus above the high stratus.

may be seen in isolated groups in fair weather conditions. But when they occur in long regular bands of 'mares' tails', or are associated with cirrostratus, they are usually caused by warm air moving high up over a mass of cold air, far in advance of a warm front (p.37) and so presage bad weather.

Cirrostratus is a thin sheet of ice crystals covering the sky. Instead of a deep blue, there is a hazy or milky appearance. A halo may be seen around the sun or moon, as light is refracted through the crystals. Such clouds are also likely to have been formed in warmer air from moisture ascending over a mass of cold air in advance of a warm front. Precipitation is likely to follow as this layer thickens.

Cirrocumulus. If the warm air overriding the cold air is unstable, small globular cloud masses may be formed at this height, showing vertical development. They are often in lines, giving the appearance of a 'mackerel sky', and hinting at unsettled conditions.

3.9 Orographic effects – forced ascent due to relief barriers

When landforms such as mountains, scarps, or plateau edges deflect air upward, condensation and precipitation may follow, heavy if the air is unstable.

The amount of precipitation depends on the direction of movement, moisture content, and stability of the air; also on local relief, such as steep slopes, re-entrants, funnels, and other topographical features. This is illustrated on a large scale in north-east India, where parts of the Khasi hills funnel humid monsoon air upward, and Cherrapunji receives an average of over 7500 mm of rain during the three wettest months; more than 1000 mm has fallen **in a single day**. Monsoon air sweeping up India's Western Ghats brings an annual fall of over 5000 mm to some seaward-facing heights, yet lee slopes a few kilometres inland have less than 750 mm.

The movement of air over very large barriers is apt to result in more than a simple rain-shadow effect. Fig. 3.15 shows that low pressure in the lee of a mountain range causes eddying and wave formation, with rolls of lens-shaped clouds maintained in the wake.

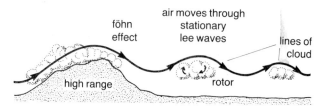

Fig. 3.15 Lee waves and cloud formation.

A flow of moist air over such mountains may also create a **föhn effect**. Pressure differences over areas flanking the mountains may cause air to move up and over the main ranges. As moist, but unsaturated, air moves up the windward slopes, it cools adiabatically by about 1 C° per 100 metres. At condensation level the water vapour forms droplets and releases latent heat. It continues to rise and cool at the slower saturated adiabatic rate, about 0.5 C° per 100 metres. Fig. 3.17 assumes that the SALR is constant, and shows temperatures at intervals of 1000 metres.

As air descends the lee slopes it gains heat at the dry adiabatic rate; so it now warms relatively rapidly. At equivalent altitudes it has a higher temperature than it had on ascent. Thus a wind may blow away from the lee slope with a temperature 10–20 C° higher than the air on the windward side, sometimes more. Its absolute water vapour content is reduced; so, with the higher temperature, its relative humidity is very low.

In fact, other processes, involving turbulence and eddying on the lee slopes, may cause a forced descent of air, so that sometimes the temperature may increase on the lee flanks though no condensation occurs on the windward slopes.

Fig. 3.16 Lee waves, formed as moist air moves up and over India's Western Ghats onto the dry Deccan plateau, are indicated by long lines of cloud alternating with open sky between; they form a stationary pattern.

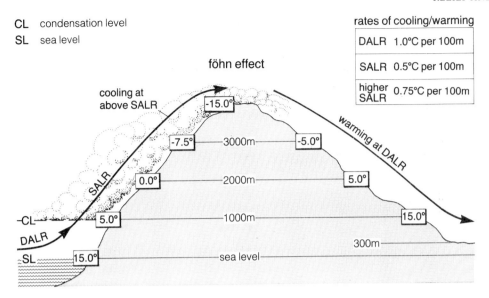

CL   condensation level
SL   sea level

rates of cooling/warming

DALR	1.0°C per 100m
SALR	0.5°C per 100m
higher SALR	0.75°C per 100m

föhn effect

cooling at above SALR

-15.0°

-7.5°   3000m   -5.0°

0.0°   2000m   5.0°

5.0°   1000m   15.0°

SALR

DALR

CL

SL   15.0°   sea level   300m

warming at DALR

Fig. 3.17 The mechanism of a föhn wind. An induced flow across a mountain system usually causes turbulent movements as sinking air streams away leeward of the barrier.

Föhn winds take their name from the winds which occur in the European Alps in spring and autumn. Air drawn over the Alps from the south usually produces marked temperature contrasts. Such winds set in when a depression lies over north-central Europe, so that a gradient is established from the Mediterranean towards the low pressure to the north. The onset of a föhn wind, with rapid thawing, is apt to trigger avalanches.

Similar winds occur in other mountain systems under appropriate pressure conditions. The Chinook wind of the Rockies, like the föhn, tends to cause rapid snow-melt in the spring. It helps to expose prairie pastures and provide moisture. The föhn winds of the Southern Alps in New Zealand have a dessicating effect on farmlands in the Canterbury plains, where windbreaks are familiar features.

3.10 Rainfall intensity

The intensity of rainfall during a storm, that is the amount divided by the duration, can have dramatic effects on physical landforms and settlements. Rapid run-off, as sheetwash, can remove and transport both top-soil and regolith; gullying may be intensified, destroying farmland; and sudden flooding can be a hazard, though subsequent deposition of alluvium may be beneficial.

In downpours, high intensity rainfall is associated with **increased drop size** rather than an increased number of drops. The average intensity for short periods is much greater than for long ones. In north-western Jamaica 200 mm of rain was recorded in a fifteen-minute storm.

Fig. 3.18 Ragged cloud, beneath higher altostratus, circulating a storm centre. In the aftermath of a cyclonic downpour, which gave parts of eastern Queensland 300 mm of rain, water poured through deep channels cut into land cleared for pasture.

CHAPTER

4

ATMOSPHERIC CIRCULATION

4.1 Energy balance in the earth–atmosphere system

There is an overall balance between the heat energy received by the earth and atmosphere and that lost to space. However, there are numerous energy transfers within the atmosphere and the conditions in the troposphere are far from uniform.

As we have seen, the insolation received varies with latitude. In the low latitudes, where the midday sun is high in the sky and insolation intense, the annual amount of incoming radiant energy exceeds that lost. But in the high and middle latitudes, where the sun's rays are always oblique to the surface and insolation less intense, annually more energy is lost than received, despite the many hours of insolation during summer (Fig. 4.1).

If, therefore, the low latitudes are not to become hotter and hotter and the high latitudes cooler and cooler, some of the *heat energy must be transferred poleward*.

The unequal heating of the earth's surface is not

a matter of gradual change from low to high latitudes. The location of the hottest parts of the earth varies with the seasons, as do the influences of the land and water masses, which absorb and radiate heat at different rates. There are also innumerable variations in surface texture and topography which affect air temperatures and energy transfers in the lower troposphere.

4.2 Seasonal variations in mean monthly air temperatures

Figs 4.2 and 4.3 show mean monthly air temperatures for January and July. In general the east-west trend of the isotherms reflects the overall decrease in insolation received from equatorial to polar latitudes. They are more nearly parallel to one another in the southern hemisphere, where ocean expanses form a more uniform surface. Their wide spacing in the tropics emphasises that over large parts of the earth's surface variations in mean air temperatures are minimal. Compare this with the lower middle latitudes, and also notice that the spacing is wider again in polar regions.

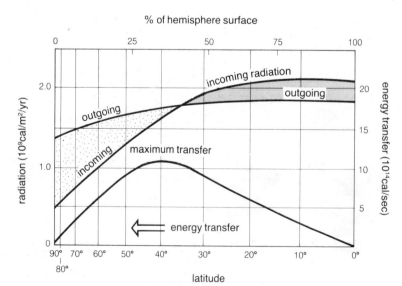

Fig. 4.1 The energy imbalance, and the mean meridional transfers which are affected by air and water circulations.

The effects of the physical differences of oceans and landmasses and the influences of oceanic circulation, which also transport heat energy, are very noticeable. The east-west run of the isotherms changes where air modified by the ocean meets that over the continent. Sharp deviations indicate warm or cold currents in offshore waters.

The January map reflects the overall heat losses from the northern landmasses. The effects of the relatively warm waters of the northern oceans on air temperatures are demonstrated by the poleward trends of the isotherms. Oceanic influences are also shown by the different mean air temperatures to the east and west of both southern Africa and South America, where there are cold currents

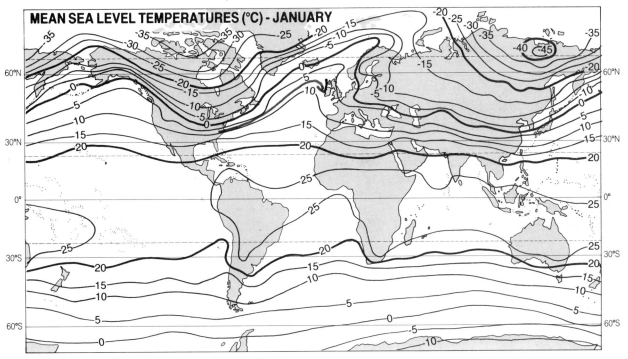

Fig. 4.2 Notice the northern oceanic–continental influences and the contrasts between the northern 'land' and southern 'oceanic' hemispheres.

Fig. 4.3 Compare the temperatures in the northern interiors with those in January.

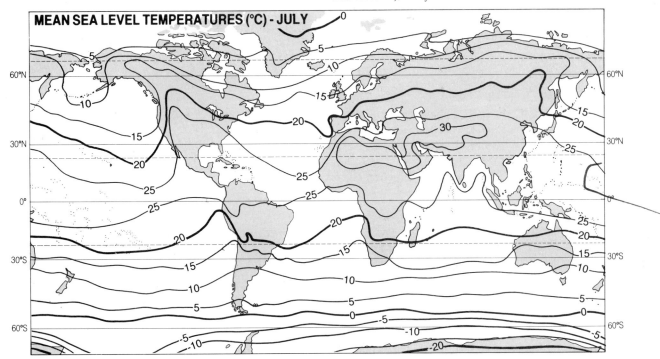

to the east and warm ones to the west. The tropical parts of the southern continents have particularly high mean temperatures at this time of year.

The July map shows that the interiors of the northern landmasses are now remarkably hot compared with conditions in winter. The northern tropical and sub-tropical desert regions, influenced by air subsidence, are very hot indeed. But, worldwide, the contrasts of mean air temperatures over both oceans and land areas are less than in January. The southern hemisphere does not show the marked contrasts between summers and winters as the northern, but there is a general latitudinal temperature shift, shown by the isotherms, from one season to another.

A comparison of the two maps gives an idea of the mean annual temperature ranges. Notice the large annual range of the northern continental interiors and the small annual range throughout the southern hemisphere. This is clearly illustrated by the maps of anomalies (Figs 4.4 and 4.5).

4.3   Isotherm maps: uses and limitations

It is essential to understand exactly what the figures plotted on isotherm maps refer to – what they tell us and what they conceal.

1 Maps of isotherms based on actual shade temperatures, recorded at surface level in a Stevenson screen, often resemble relief maps. But the use of **sea-level isotherms**, based on a calculation of what the temperature would have been at sea level, have the virtue of eliminating some of the relief effects and highlighting influences such as a prevailing wind, ocean current, or 'continentality'.

2 A **mean daily temperature** is calculated from the day's maximum and minimum readings by adding them and dividing by two. But a daily mean of 27°C may conceal a maximum of 40°C and minimum of 14°C, or a maximum of 32°C and a minimum of 22°C. It will also mask small local peculiarities during the day, perhaps a sudden fall in temperature which might be vital to a certain crop.

A **mean monthly temperature** for, say, July is obtained by adding the mean temperatures for all days in July and dividing by 31. But a monthly mean of, say, 20°C may well hide a hot spell when daily temperatures soared to 30°C, or a cool spell when they failed to reach 15°C. As the mean July temperature for that place is the average of 35 consecutive July mean figures, the limitations of such a figure are obvious.

In fact the usefulness of any mean figure varies with location. If the weather is variable, as in the cool temperate oceanic regions, a large number of observations are needed. Whereas, for a place in the low latitudes, fewer readings may give reasonably reliable mean temperature figures.

3 The **annual temperature range** is the difference between the mean temperatures of the hottest and coldest months. This may be far from the **absolute temperature range** in that year. A mean of 5°C for the coldest month may hide an actual temperature of −15°C, and 15°C for the hottest may hide a temperature of 35°C. In this case the annual temperature range would be 10 C°, but the absolute temperature range 50 C°.

4 Besides these particular yearly ranges, the **mean annual range** and the **mean absolute maximum and minimum** figures may be obtained. The latter may be a more significant figure for engineers and farmers than the mean annual range. They may also be interested in figures showing the **mean duration** or **frequency of occurrence** of certain temperatures.

5 Many other isothermal maps may be drawn to portray required climatic information, such as those showing **variation from the mean temperature for a given latitude** on a worldwide basis. Notice, for instance, how the influences of the cold currents to the west of the continents can be seen in Figs 4.4 and 4.5.

4.4   Energy transfer and air circulation

About 60 per cent of the energy transferred from lower to higher latitudes takes place through circulations of air in the lower and upper troposphere. Ocean currents, whose movements are mainly caused by wind drag, transfer about 25 per cent. Variations in sea-water density also create slow circulations between surface and deep ocean waters, between polar regions and the tropics, but play only a minor role in heat transfer.

Apart from the kinetic energy transferred by virtue of its movements, air acquires water vapour by evaporation from the oceans; this releases latent heat on condensation under cooler conditions.

Global air circulation is not simply a matter of air movements created by latitudinal temperature differences, with two convection cells extending from the equator to the poles. The composition and properties of the atmosphere vary from place to place. Masses of air acquire different properties as they move, and so are heated or cooled, or moistened. Consequent changes in density and relative humidity may cause strong vertical movements, which set in motion related air flows in the high troposphere and horizontal winds near the surface. The whole atmosphere moves with the earth, so all such movements are influenced by the Coriolis force (p.6), by friction at the surface, and by the resulting Ekman spiral effect.

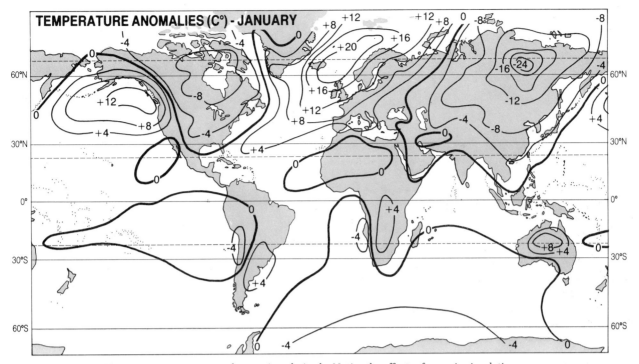

Fig. 4.4   Variations from mean temperatures along a given latitude. Notice the effects of oceanic circulations.

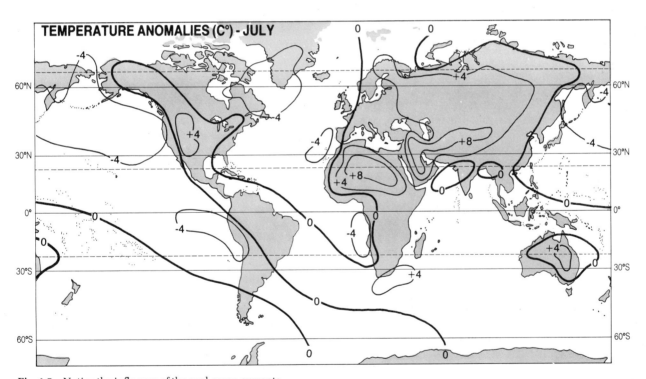

Fig. 4.5   Notice the influence of the cool ocean currents.

4.5   A simple pattern of circulation

We still have much to discover about the mechanisms of the complex, changing atmospheric circulation. However, certain atmospheric movements occur regularly enough for us to recognise patterns of air pressure distribution and winds, which are revealed by seasonal maps such as Figs 4.7 and 4.8.

As long ago as 1735, George Hadley suggested that a global convection system was set in motion by a tropical heat source. Then in 1856 William Ferrel put forward a circulation pattern in each hemisphere, based on the three cells shown in Fig. 4.6. This model is a great over-simplification, but acts as a useful starting point from which to

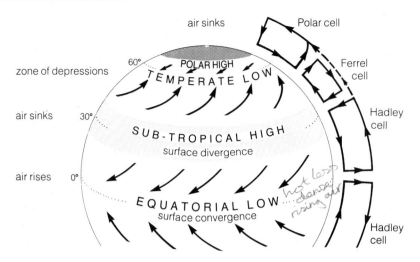

Fig. 4.6 A simplified view of zonal air pressure and surface winds, related to three simple cells.

focus on more complex vertical and horizontal air movements.

In each hemisphere he recognised the **Hadley cell** as a convection circulation, with air rising in equatorial regions and moving poleward at high altitudes, though in fact deflected as a westerly flow. Air sinks in the vicinity of latitude 30°, so that at the surface there is divergence. Some of the air moves back to the low latitudes, deflected as it moves to form the easterly flow of the Trade winds; other air moves poleward as surface westerlies.

A **polar cell** was seen as a response to cold, dense air sinking in polar regions and returning towards lower latitudes as easterly winds. Between the Hadley cell and the polar cell, on the original model, was the **Ferrel cell**; a cell which arose in response to the movements set up by the other two.

These cellular circulations seemed to explain the presence of areas with considerable precipitation, where air in general is rising, and dry zones, notably the world's hot deserts, where the air is sinking, establishing the sub-tropical zones of high pressure.

4.6   Sea-level air pressure and mean seasonal winds

Before taking a more realistic view of global air circulation and the causes of existing weather patterns, a glance at Figs 4.7 and 4.8 shows that, though the continuous belts of high and low pressure of the model do not exist, average seasonal distributions of pressure and wind directions accord in many respects with those of the model.

A broad zone of low pressure extends across the low latitudes, into which the tropical easterlies appear to converge. Over the sub-tropics there are well-defined belts of high pressure, especially in the summer hemisphere. On the maps of average pressure, however, they appear as strong cells over the oceans, extending over land areas.

Poleward of these high-pressure belts are low pressure zones, particularly well-defined in the southern hemisphere, where the expanses of ocean are interrupted by only relatively narrow landmasses. Between latitudes 35°–60° mid-latitude westerlies flow across the pressure gradients, as the so-called 'roaring forties' and 'furious fifties'.

In the northern hemisphere the westerly flows are affected by the landmasses and the thermally-induced pressure conditions associated with them. The topographical barriers, friction effects, and high temperature ranges of the land areas contrast with the oceans' surface uniformity and their ability to act as a heat reservoir. This is reflected in the marked pressure differences over the continents and oceans in each season.

In polar latitudes, high pressure, mainly thermally-induced, persists over Antarctica for most of the year. But the air pressure is more variable in the northern polar lands, and the Arctic highs are less persistent. The polar easterlies are hardly regular winds; but irregular outward flows of polar air greatly influence weather conditions in the higher middle latitudes.

The thermal controls of the landmasses are also seen in the summer monsoonal inflows of air, especially in southern and eastern Asia, and in the outflows from central Asia during winter.

Clearly such maps of average conditions mask very variable weather in many parts of the world; notably in the mid-latitude westerlies zone, where eastward-moving depressions are accompanied by day-to-day variations in wind direction. Even the apparently persistent zones and cells of high pressure of the southern hemisphere are not stationary systems: they are a procession of highs, separated by troughs of lower pressure, moving from west to east.

Air also accumulates in certain source regions, and remains long enough in contact with the local surface for properties of energy and humidity to be acquired and transmitted through its mass.

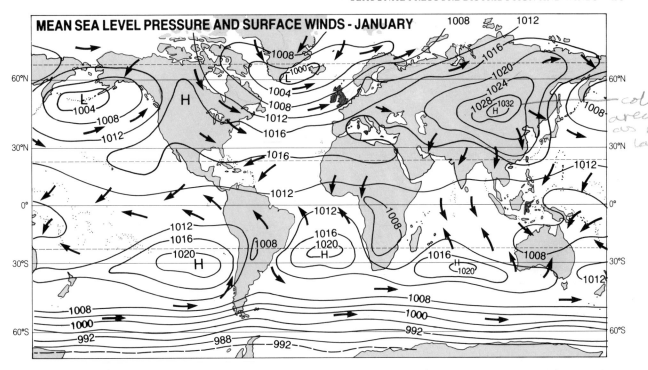

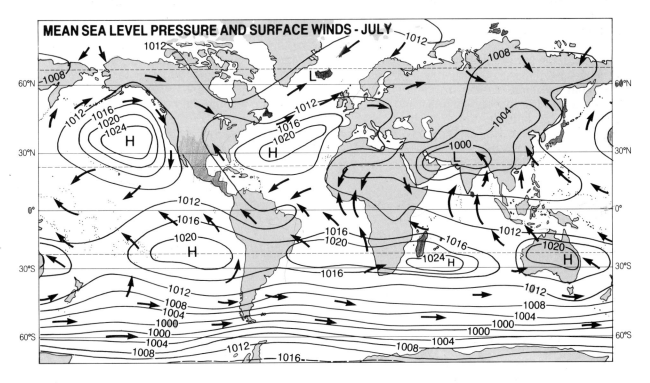

Fig. 4.7 The patterns of the temperate westerlies, Trade winds and monsoon flows related to monthly average pressures.

Fig. 4.8 Pressure conditions which highlight the effects of the Asiatic monsoons.

Such air masses (p.39) move outward from their source in response to pressure gradients and affect the weather of the regions they invade.

However, although these maps of average pressure conditions, and of the winds responding to the average gradients, tell little of the properties of moving air or of changing weather, they do give a generalised picture of air circulation. This makes them useful stepping stones towards understanding energy distribution and world climates.

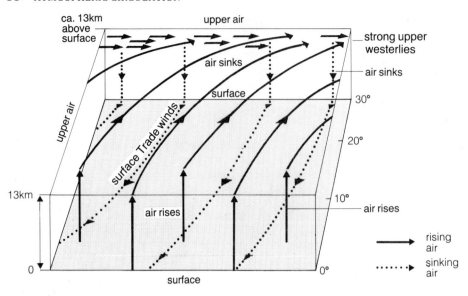

Fig. 4.9 The Hadley cell and the upper westerlies.

4.7   Upper air and the global circulation

The Hadley cell is undoubtedly a means of transferring heat from the low latitudes, though the mechanism is rather more complex than Hadley envisaged. There *are* large inflows of surface air and outflows in the upper atmosphere. At centres of convection across the low latitudes the energy of the air rising from heated surfaces (in response to intense insolation) is increased by latent heat released by condensation in numerous cumulonimbus clouds (Fig. 4.26).

The high-level outflows form the north-south/south-north component of the Hadley cells, though there are also west-east upper air movements.

The earth's circumference at the equator is approximately 40 000 km, but at latitude 30° the east-west (zonal) distance around the earth is only 35 000 km. As air moves poleward in the upper troposphere, in the Hadley cell, it conserves its angular momentum (mass × rate of motion through 360° about the earth's axis), and so increases its speed relative to the earth's surface.

There is very little friction in the upper atmosphere and the Coriolis force increases away from

Fig. 4.10 A generalised picture of the strong upper westerly flows and the core areas, with powerful jet streams.

JANUARY

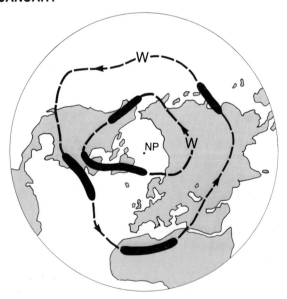

JULY

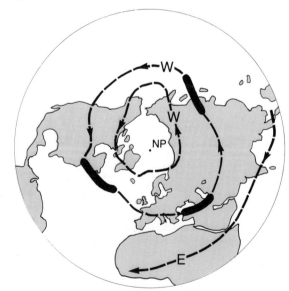

– – ▶ – – mean axis of strong upper air flows [in reality paths and zonal index of upper westerlies are variable]

▬▬▬ core area of upper winds of exceptional speed (80-180knots) [minor jet streams develop occasionally]

AIR TEMPERATURE mid summer

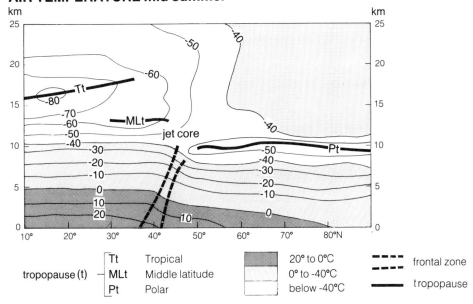

Fig. 4.11 Air temperatures during the northern summer at various altitudes. Notice the meridional temperature differences, which reflect the need for energy exchanges, and the jet stream as the core of the strong upper westerlies. Variations can be seen in the height of the tropopause.

the equator: so as the air travels poleward it turns eastward as a geostrophic wind of increasing speed. However, the upper air only travels in a poleward direction between the equator and about 25°–35° latitude; then the wind becomes westerly, part of strong **upper westerlies** typical of these latitudes.

At about latitude 30° the core of these powerful westerly air currents is particularly concentrated over some parts of the globe, forming what is known as **the sub-tropical jet stream**, some hun-

dreds of kilometres wide and several kilometres deep (Fig. 4.10). There seems to be a limit to the speed of the jet stream, for turbulence and energy losses keep it to the order of 100–150 m sec^{-1}, and usually below this.

Beneath the jet stream the air subsides, creating zones of high pressure at the surface. High-level air feeds into the jet stream, but, as speeds are limited, some sinks to the surface. At one time it was thought that the sinking was due to air cooling as it moved poleward; then to a piling up

PRESSURE GRADIENTS

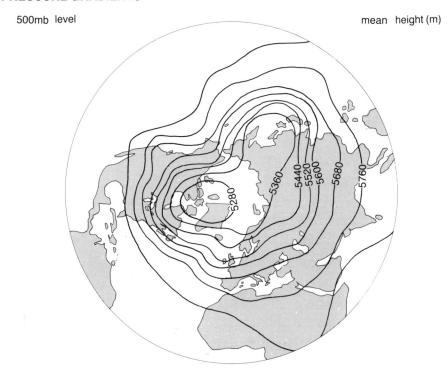

Fig. 4.12 The mean heights of the 500 mb level and the gradients involved indicate that geostrophic winds will produce a looping circumpolar pattern, as strong upper westerlies.

of the poleward-moving air at about latitude 30°. There are other reasons, however, why the air should sink. In any meandering jet stream, the air moving equatorwards in one arm of a loop slows down. This causes upper air convergence and generates a cell of sinking air, so that high pressure (anticyclonic) conditions are established below.

The velocity of the strong upper westerlies varies as they continue around the globe. Most of the transfer of heat takes place at the westward end of sub-tropical cells of high pressure. Over Asia the jet stream has a less continuous flow during the northern summer. In fact high pressure develops far above the heated Tibetan plateau and a high-level southerly flow develops in the upper air. This is deflected to form **upper easterlies**, which influence weather conditions from southeast Asia across to West Africa. (p.43).

The world's hot deserts lie beneath the sinking air of the sub-tropics, for such subsidence effectively blankets most of the columns of air rising by convection from the hot surface.

The sinking air of the sub-tropics diverges at the surface. The easterly Trade winds complete the Hadley cell as they return to low latitudes. The surface air also moves poleward. As it does so it transfers heat energy by irregular movements and circulations, rather than as a simple flow of westerly air.

The **mid-latitude air circulation** is highly complex. In places the air builds up, creating areas of high pressure (anticyclones) from which it flows slowly outward; while low pressure cyclonic depressions, with air spiralling strongly inward, move in a general easterly direction across these latitudes, especially over the oceans. They are closely related to pressure systems and air movements in the upper atmosphere.

In the atmosphere in the middle latitudes, north-south temperature contrasts are much greater than in the tropics. Pressure gradients in response to temperature differences cause the upper air at first to flow poleward; but the Coriolis force,

which increases towards the poles, deflects these high-level winds towards the east. Any slackening of speed, with air convergence, favours subsidence; acceleration, with air divergence, draws up air from below.

Subsidence over polar regions tends to cause outward surges of cold surface air. Fig. 4.13 shows its clash with warmer air of sub-tropical or tropical origin as a **polar front**, with a horizontal temperature gradient extending right up through the troposphere. But such fronts are not continuous, nor are they always zonal (running east–west).

AIR SPEED mid-summer
(knots)

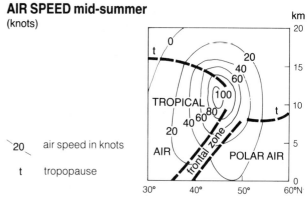

Fig. 4.13 The core of the upper westerlies in July.

The **upper westerlies** form a looping stream of fast-moving air. At the core of these strong winds are **polar front jet streams**. Fig. 4.14 indicates the mean seasonal positions of the upper westerlies and the mean locations of the most concentrated air flows. But the actual paths of the upper westerlies vary, as does the velocity of the air in the jet streams.

A series of long waves, the **Rossby waves**, form in the westerlies as they encircle the polar regions. There are usually between three and five of them in the northern hemisphere, while three is more usual in the southern hemisphere. Smaller waves travel through them. These are closely connected with pressure changes at the surface.

MEAN WIND SPEEDS (knots)

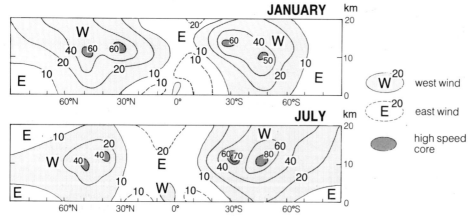

Fig. 4.14 The meridional energy exchanges result in persistent circumpolar westerly winds in each hemisphere. Concentration into jet streams shows seasonal latitudinal variations. In the low latitudes the strength of the upper easterlies increases during the northern summer.

High physical barriers, such as the Rocky Mountains, affect the upper air wind-speed and wave amplitude. They tend to perpetuate the pattern of loops formed in the long waves as they travel eastward, for as an airstream rises over a high barrier it contracts vertically. This decreases its speed of rotation (vorticity), much as raising the arms decelerates a skater's spin. But the earth's vorticity increases toward the poles; so in the northern hemisphere the decelerating air tends to move equatorward with a clockwise (anticyclonic) movement. As the airstream descends in the lee of the barrier, the air column is stretched vertically, increasing the rate of spin; so it moves poleward curving cyclonically. The waves thus set in motion are perpetuated downstream. Fig. 4.15 points to a well-marked wave pattern and concentrated air flow over a relatively small area above eastern USA.

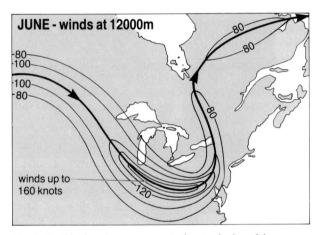

Fig. 4.15 The looping upper westerlies in the lee of the western mountains, with strong jet winds about latitude 40°N.

The amplitude of the waves tends to increase or decrease as energy is gained or lost, and the speed of flow varies accordingly. A strong westerly flow retains its zonal (west-east movements) with only gentle looping. It is said to have a **high zonal index**. This is a common winter feature, when the temperature contrasts are greater and the circulation strong. At other times there may be considerable looping, and so more north–south/south–north (meridional) movements of the air stream. As energy inputs, strong or weak, have an important influence on the speed of the upper westerlies, the location of warm and cold ocean currents and the changes in heat sources in the continental interiors affect the wave pattern. This in turn affects the development of cyclones and anticyclones at the surface.

4.8 Upper air conditions and weather at the surface

Weather sequences at the earth's surface and the nature of the upper air flows are closely interrelated. High aloft there are ridges of high pressure and troughs of low pressure associated with the waves in the upper westerlies. Upwind of a ridge the air flows poleward, in an anticyclonic movement; so, boosted by centrifugal force, it speeds up. But as this turns to flow equatorward it slows, and so the air piles up; that is **convergence** takes place. As it curves into a cyclonic flow, the centrifugal force acts against the gradient force. Then, as the air speeds up again as a poleward flow, there is **divergence**.

The convergence in the eastern arm of a ridge causes air to descend and create anticyclones at the surface. Divergence in the western limb of the

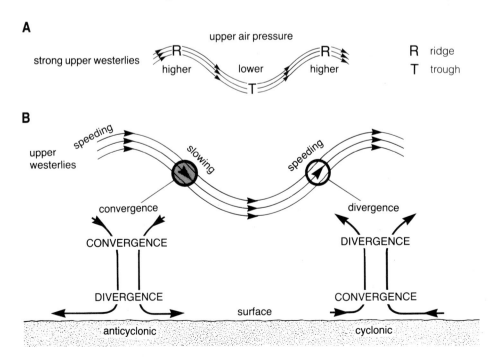

Fig. 4.16 A shows loops in the upper westerlies related to air pressure differences which create ridges and troughs. **B** shows how slowing produces convergence and sinking, while acceleration draws in air from below.

high zonal index

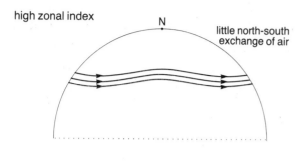

low zonal index

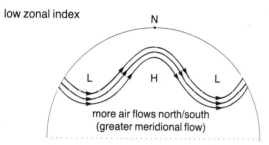

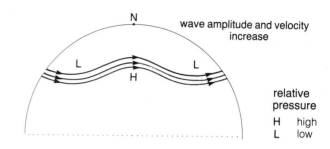

relative
pressure

H high
L low

break-up

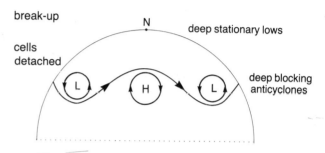

Fig. 4.17 The strong zonal flow, with few ridges or troughs, does not make for the establishment of deep lows or blocking highs of the kind that develop as the amplitude of the loops increases. 4.17 A–D show near-surface flows corresponding to upper conditions.

ridge allows near-surface air to ascend, creating low pressure below, with surface convergence, favouring the development of cyclonic systems.

In fact the energy changes between surface air and upper air are complex. The vertical exchanges are not all initiated by the upper air conditions. Energy is injected into the upper air by strong convective updraughts and by upward movements associated with frontal clashes between airmasses of different temperature. As we have seen, this can affect the amplitude of the Rossby waves and create irregular variations in the zonal index of the upper westerlies, which may last for a month or two.

Fig. 4.17 shows what happens when a strong zonal flow gives way to loops and cells and a **low zonal index**. This leads to marked meridional

flows, which strengthen the contrasts between ridges and troughs, and so favour the development of strong cyclones and anticyclones at the surface. There is a tendency for the highs so formed to drift southward and for the lows to move northward. Thus the changing weather conditions are much affected by the energy exchanged between the upper and lower parts of the atmosphere.

The following maps show **weather conditions associated with upper air flows** with a high zonal index (Figs 4.17A and 4.17B) and, by contrast, very different surface conditions corresponding to loops in the upper air streams and meridional flows at some 5000–5500 metres above the surface (Figs 4.17C and 4.17D).

Imagine the geostrophic winds in response to the lows in the upper atmosphere, and picture the general pattern of the upper air streams in each case.

In Fig. 4.17B notice the procession of lows moving eastward across the Atlantic, and the depression which brought strong winds and

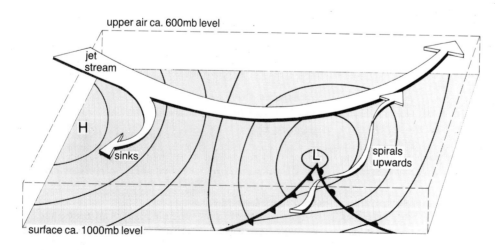

Fig. 4.18 A diagrammatic view of the relationships between high-level air convergence and divergence in the upper westerlies and the strengthening anticyclones and deepening depressions near the surface.

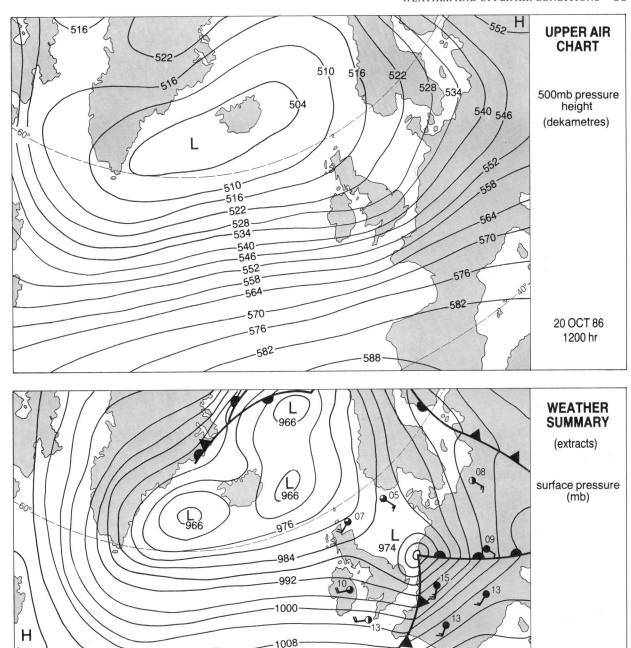

Figs 4.17 A and B   High zonal index.

heavy rain to southern England and Wales, where, after a muggy period, it has turned colder.

In Fig. 4.17D cold air has advanced from the high created over Scandinavia, pushing under warmer air and bringing sleet and snow showers to eastern England, with rain in the south-west. Notice the separations into areas of high and low pressure, and the changes taking place over Spain.

4.9   The mid-latitude depressions

In a mid-latitude cyclone, the air does not necessarily form a complete revolving wind system, so that it is better to use the term 'depression' rather than 'cyclone'.

In **cyclonic systems** the air flows in response to a well-developed pressure gradient, towards a

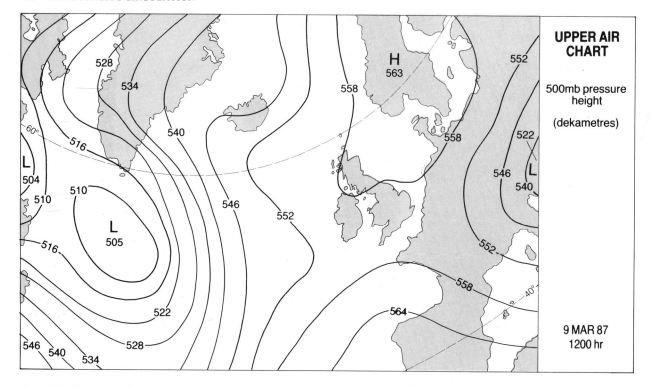

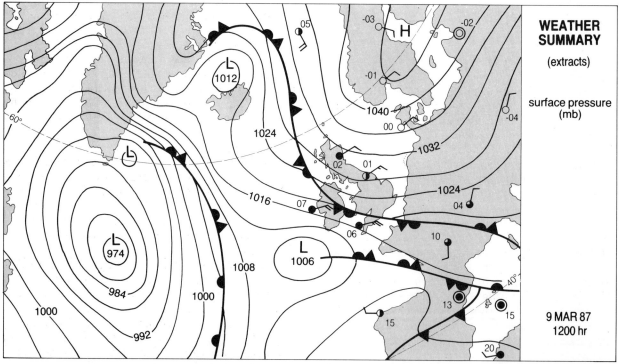

Figs 4.17 C and D Low zonal index.

centre of low pressure. But it is subject to the Coriolis force, the force of friction, and to some extent to centrifugal force; so that in the northern hemisphere the air is deflected to form an anti-clockwise swirl about the low pressure centre, and a clockwise one in the southern hemisphere.

At the centre of a well-developed depression the

air rises, much as shown in the simplified diagram (Fig. 4.18), and is drawn into the high-level flows by divergence in the upper atmosphere.

Mid-latitude depressions mostly form along a front of contact between cold, dense air of polar origin and warmer, less dense tropical maritime air. The angle of contact varies with the general

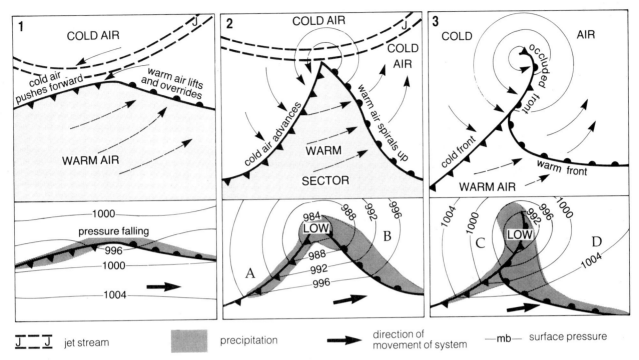

Fig. 4.19 Upper air flows and a deepening depression, leading to the gradual occlusion of warm surface air.

pressure conditions, temperature differences and speeds of movement of the masses. Its tendency to rotate also varies with the earth's vorticity, and so with latitude.

A **front** of this kind is unstable, so any kind of disturbance may cause a wave to develop along it. This sometimes happens when cold air surges into the warm air at the surface, as a cold front, so that the displaced, lighter air rises above it. Should this coincide with divergent air high above, the warm air continues to ascend.

At the surface the air begins to spiral inward towards a developing centre of low pressure. As a result, warm air moves forward, as a **warm sector** between colder, denser masses. The surface edge of this advancing warm air becomes the **warm front**. Behind the warm sector the trailing edge of the advancing cold air becomes the **cold front**. The system as a whole tends to move eastward.

Above the warm front the warm air moves up a gentle, retreating slope of polar air, and may continue upwards, towards the high westerlies. At the rear of the depression the cold air is sinking, and may be fed by convergent air in the up-wind arm of the loop of the jet stream. The cold front usually advances rather more quickly than the warm front, lifting warm air as it does so, and narrowing the warm sector.

If the air spiralling upward continues to be removed, the depression deepens as it advances. The whole system may grow to affect an area of over a thousand kilometres across.

Fig. 4.19 (2) represents a fully developed depression. Notice how the isobars show the press-

ure differences in the warm and cold air, and the sharp angle of the isobars across the front. Depressions tend to move forward in the direction of the isobars in the warm sector.

As the fronts advance and move anticlockwise, the cold air behind the system continues to encroach on the warm sector. Finally it may occlude the warm air, lifting it clear of the ground (Fig. 4.20). This state of **occlusion** begins where the warm sector is narrowest and develops outward from the low pressure centre.

Sometimes secondary waves develop on the cold front at the rear of the depression and a whole family of **secondary depressions** may circle slowly around the primary one. In the British Isles a succession of secondary depressions, with belts of rain, give long spells of changeable weather; the high pressure ridges between the lows give a drier day or two.

In some depressions, especially those travelling about an anticyclone, there is little temperature difference between the warmer and cooler air. When the air is generally subsiding, the clouds lack vertical development, so precipitation is slight. Such fronts are referred to as **kata fronts**, in contrast to **ana fronts** where warm air rises strongly, as in the usual vigorous depression.

4.10 The mid-latitude depression: a ground-level view

The first signs of an approaching depression are likely to be high feathery 'mare's tail' cirrus, where, perhaps several hundred kilometres ahead of the front, the air still retains enough moisture after ascent to form ice crystals at 8000–10 000 metres.

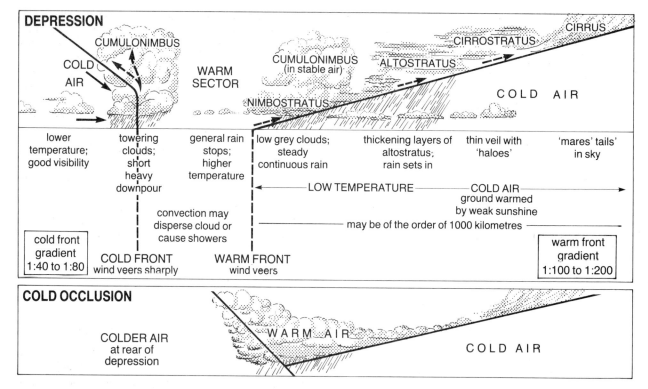

Fig. 4.20 (Above): Cloud formation and weather features associated with a well-developed mid-latitude depression. **(Below):** Inversion, with warmer air overlying the denser colder air which has advanced at the rear of the depression.

As the warm front approaches there may be a sun halo in the thin veil of cirrostratus which follows. The watery blue gives way to thicker altostratus, and spots of rain herald the rain belt ahead of the warm front. Steady persistent rain sets in as dark nimbostratus masses, with their base several hundred metres up, come in ahead of the front itself.

The intensity of the rain and the form of the clouds depend on the nature of the warm air. Relatively dry air may give little precipitation; but moist unstable air is likely to be signalled first by a 'mackerel' sky, with lines of cirrocumulus, and later by heavy rain, as vertical development produces cumulonimbus.

As the warm front passes the wind veers and the temperature rises. The barometer ceases to fall, or falls more slowly. The continuous rain stops, though the humidity remains high and there may be a shallow layer of cloud in the warm sector, or scattered convection showers, or even short periods of rain in hilly country.

Fig. 4.21 Brighter, colder air follows the billowing cumulus and torrential rain at the cold front of a depression which has passed over Vancouver from the west.

At the cold front the warm air is forced upward by the advancing cold air. The slope of the surface of discontinuity is greater than at the warm front, being of the order of 1:25 to 1:80, rather than between 1:100 and 1:400. Here the vigorous uplift of unstable air produces huge cumulonimbus clouds, usually with a relatively brief torrential downpour. Sometimes the cold air advances so rapidly that it overruns the warmer air, so that brief violent **line squalls**, with very strong vertical movements develop.

As the cold front passes, with its 'clearing-up shower', the wind veers sharply, the barometer rises, the skies clear, visibility improves, and the temperature falls as cold air moves in from a poleward direction – air which is usually colder than that at the front of the depression.

4.11 A summary of the meridional circulation

Fig. 4.22 shows a schematic view of the meridional circulation. The Hadley cell behaves much as suggested by Fig. 4.6. Air rises in the low latitudes, with much vertical heat transport effected in huge cumulonimbus clouds. In the sub-tropics subsiding air is fed by fast-moving upper air streams, thus forming and regenerating anticyclones to the west of the continents, with high pressure extending over the eastern parts of the oceans. Air returns equatorwards as the Trade winds, and rises again in the low latitude zone of convergence.

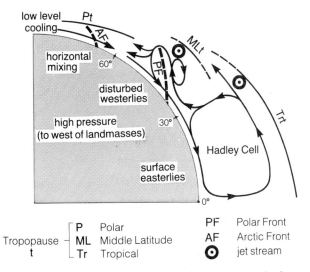

Fig. 4.22 A schematic view of air circulations between high and low latitudes.

At the eastern end of the great sub-tropical anticyclones the air is less stable, especially during the summer months. The eastern parts of the landmasses and western parts of the oceans have ample precipitation.

Polewards, the so-called Ferrel cell is better described as a broad zone in which wave cyclones are active along individual polar fronts. Fluctuations in the swing of the upper westerlies, the development of blocking highs, and the winter thermal highs in continental interiors make this a zone with variable weather conditions. In the southern hemisphere the absence of broad landmasses in the higher middle latitudes makes for strong circumpolar westerly flows.

Much of the polar air is warmed during its journey to lower latitudes. Some becomes mixed with warmer air in cyclonic systems and rises. It may thus form part of the upper westerlies and once again be transferred poleward.

An Arctic front is shown; though, in fact, this occurs most frequently in Antarctica, where cold, dry polar air meets other polar air which in circulation has been modified by the relatively warm ocean surface (p.47).

4.12 Air masses and their source regions

An air mass is a portion of the atmosphere of considerable depth whose characteristics of temperature and humidity are remarkably uniform in a horizontal direction at any level, at or above the earth's surface.

In certain parts of the world air accumulates and remains stationary at its place of origin long enough to acquire specific characteristics. The air is conditioned from the earth's surface; so the properties of warmth or coldness, humidity or dryness, become strikingly homogeneous when the air rests over an extensive uniform area.

Its stability depends mostly on the vertical temperature distribution – temperatures at various levels in the mass. If the air remains over a particular surface for a long time, any vertical movements will help to distribute surface characteristics throughout the mass.

There are certain source regions where air masses develop typical properties, particularly where anticyclonic conditions persist for long periods, as over the eastern parts of the oceans in the subtropics, the Antarctic icefields, and the cold interiors of Euro-Asia and northern North America during winter. Here air subsides, and diverges at the surface with light movements which allow uniform characteristics to develop. The humidity is usually very low in an air mass building over a cold land area.

As the air moves away from the source region, other surface influences effect changes. For instance, as dry air subsiding over the eastern Atlantic off the Saharan coast moves westward over the ocean, it rapidly acquires a high water-vapour content. Fig. 4.26 shows how cloud formation and storm development increase westward, as the subsidence becomes weaker and moisture from the surface is spread through the air by convection.

4.13   Movements and clashes of air masses

As air moves away from the source area in response to pressure gradients it tends to retain its characteristics for some time; though, as we have seen, it is gradually modified by the surface beneath. It may also be modified by inputs of solar radiation. Temperature and humidity changes take place slowly, unless strong vertical movements are set up.

The air mass affects the weather of the regions it invades, in ways which depend on prevailing local conditions. Warm, humid air advancing and replacing cold, dry air is likely to bring a belt of frontal rain and milder conditions; but when replacing a declining summer anticyclone, it may bring thunderstorms and cloudy, cooler weather.

The actual path of movement of an air mass responds to barometric gradients; though really outstanding relief features may influence it. Fig. 4.23 shows some of the air masses which affect the British Isles. Certain air masses are more dominant in a particular season, though the movements of these bodies of air are irregular.

The idea of a global wind system is still valid, but a knowledge of the origin and nature of air masses gives a clearer picture of atmospheric movements superimposed on the general circula-

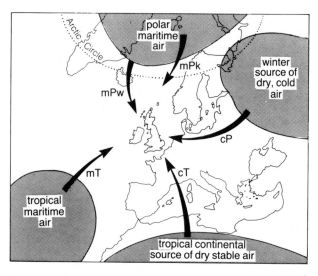

Fig. 4.23   Air mass sources which affect the British Isles. Air is likely to be modified by passage over the ocean. Often it does not arrive direct from a source region, but after circulation about a particular pressure system. Flows of mT air are the most frequent through the year; those of cP and cT air much less so.

tion. The location of air masses and their fronts are important synoptic features, plotted on weather maps. Climatically, Fig. 4.24 shows their **average** positions during January and July.

AIR MASS SOURCE REGIONS

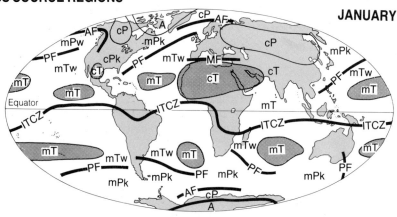

JANUARY

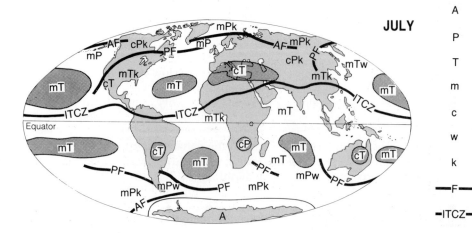

JULY

Fig. 4.24   These are the mean locations of air mass source areas in winter and summer. The fronts shown really represent narrow zones, within which air of contrasting properties meet from time to time, rather than fixed frontal positions. Similarly, the ITCZ is the mean central location of a broad zone of converging tropical air flows. Notice the influence of strong summer heating over the landmasses.

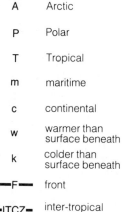

A	Arctic
P	Polar
T	Tropical
m	maritime
c	continental
w	warmer than surface beneath
k	colder than surface beneath
—F—	front
-ITCZ-	inter-tropical convergence zone

4.14 Characteristics of air masses

Arctic and Antarctic cA and mA air. cA air has its origins over the frozen surfaces and ice-caps of the very high latitudes. The moisture content of the very cold air is low; so, with inversion conditions, it is very stable. As it passes over warmer seas, heat and moisture are taken up. Contrasts are established which lead to instability and convectional movements in this mA air.

Polar cP and mP air. cP air has its origin over cold land or water in high latitudes. It is very dry and cold in its lower layers, often with strong inversion. But it, too, is readily modified by warmer seas, so that polar maritime (mP) air tends to be less stable. Sometimes an **Arctic front** is formed where very cold cP air meets warmer mP air, warmed as it rotates as part of a depression system. Fronts of this kind are mostly established about the Antarctic.

Tropical cT and mT air. Air which sinks onto the cT source regions, in lower middle or sub-tropical latitudes, remains warm and dry as long as the blanketing effect prevents convective updraughts reaching sufficient altitude to cause condensation; and so it is stable. Tropical maritime (mT) air, with moisture acquired from the ocean, may remain stable as it moves from the source region if the lower air is chilled from a cold surface; but passage over a warmer surface increases instability.

The suffix *w* is used to show that the air mass is warmer than the surface, and *k* (kalt) that it is colder, so that the symbols mTw and mTk point to likely weather conditions.

In the tropics there are not the strong temperature contrasts between air and surface that exist in temperate regions, so that maritime air often remains stable for a long time; even over the oceans

fine weather with scattered cumulus is typical of the Trade wind zones. Nevertheless, upper air divergence or forced uplift due to high relief may cause heavy rain.

As we have seen, the fronts between tropical and polar air masses in middle latitudes vary in position. Occasional surges of cold air in winter may cause cP and mT air to meet at what is sometimes called a **Mediterranean front**. But within the tropics convergence of contrasting air masses is less frequent.

4.15 The inter-tropical zone of convergence (ITCZ)

Tropical airflows from sources in opposite hemispheres meet in a broad zone. There are not the temperature and density differences that are found between conflicting air masses in higher latitudes, nor are there well-defined fronts.

However, equatorial weather is far from uniform. The warmth and humidity of converging air masses makes for potential turbulence, and the air may be triggered into strong uplift. But the main disturbances often come from waves in the air flows themselves.

Sometimes, during low zonal index conditions, an equatorward swing of a long wave in the upper westerlies may lower the upper air pressure over a low-level trough in mT air, causing its cyclonic movements to become more active, and perhaps trigger a tropical cyclone (p.49). Hurricanes and severe storms are more frequent on the eastern side of the oceans, where the mT air is less stable; though they seldom occur within five degrees of the equator.

When land surfaces of the outer tropics heat up during the summer months, mT air may be drawn in and create a **zone of instability** and a period of heavy rainfall. The occurrence of such summer precipitation is of vital importance to occupied

Fig. 4.25 One of a series of tropical rainstorms in unstable air from the western Atlantic, passing north of Antigua. Notice the anvil effect above the column of cumulus cloud. The lower clouds show a regular level of condensation and form individual billowing masses.

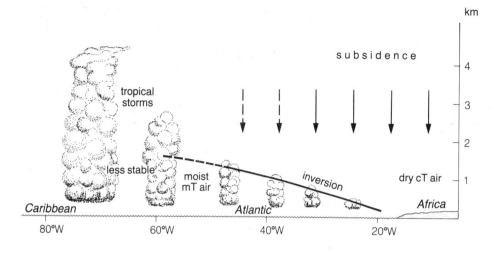

Fig. 4.26 Subsiding air in the sub-tropics keeps the eastern Atlantic dry. The Trade winds pick up moisture from the ocean; and further west the potentially unstable air is able to create lines of towering cumulus cloud masses and occasional rainstorms, as in Fig. 4.25.

parts of the arid Sahel countries, south of the Sahara. The main rain area is usually within the body of the mT air, which meets cT air at its poleward limits. But in some years cT air prevails in the Sahel zone for most of the summer, preventing the less stable, moister air from the Gulf of Guinea to advance over what soon become drought-stricken lands (Fig. 4.29). When this situation continues for years at a time, it creates widespread famine and desertification; problems which are considered further on pages 152–3.

Summer indraughts of mT air are also characteristic of the great monsoon systems.

4.16 Monsoon systems

In the monsoon areas, thermal effects cause pressure changes, so that winter and summer air flows are reversed; though the depth of atmosphere involved in the inflow is not great.

Monsoons are particularly well developed in southern and eastern Asia, and strong seasonal reversals occur over northern Australia. Those which affect the southern parts of West Africa and south-eastern USA and the Gulf states have complicating factors: in south-east USA there is not a persistent outflow of air during winter.

The monsoon in southern Asia

This is in many ways a separate system from that of eastern Asia. Here the Himalayas are a real climatic barrier.

During winter strong **upper westerlies** are established south of the Himalayas, so that air subsides over the northern plains of the Indian sub-continent. In a sense, the normal Trade wind flow is established over India. Air moves towards the inter-tropical zone of convergence, now south of the equator.

Maps of prevailing winter winds show air flowing outward, from the exceptionally high pressure areas of central Asia, over China and India. But the outflow of cold air over eastern China is different in character to that which descends, warmed by compression, over northern India.

From November to February the mean monthly temperatures of the northern plains are between 10°–16°C. The skies are clear, midday temperatures very warm, and nights crisply cool. From time to time, however, weak depressions move into north-west India from the west.

Towards the end of winter it becomes hotter, and from mid-March the midday heat becomes intense. Mean temperatures rise rapidly and pressure falls over the northern plains. For a while this area of low pressure is separated by a declining high from the very warm, humid mT air over the Indian Ocean.

By the end of May a zone of convergence becomes established in the region of the deep thermal low, so that mT air is able to advance as far as 30°N. This 'burst' of the monsoon occurs at almost the same time each year; so regularly that agricultural practices become geared to the average time of arrival. Occasional delays can cause widespread crop failures.

The sources of the low level moist inflow are in the southern hemisphere, so that mT air crosses great expanses of ocean in equatorial latitudes, and is sometimes classified as mE air.

Though moist air moves over much of the sub-continent, the idea of a simple flow bringing rain to affected areas is not sufficiently accurate. Daily conditions vary a great deal. Shallow disturbances arise in the maritime air and bring variations in the occurrence and intensity of rainfall as they move across the land.

During spring the sub-tropical jet moves northward, allowing the summer monsoon system to become established. The variable weather conditions and the amount and distribution of rainfall, partly relate to the fact that strong upper westerlies blow to the north of the Himalayas, while in early summer an **easterly jet stream** develops to the south. The Tibetan plateau is over 4000 m

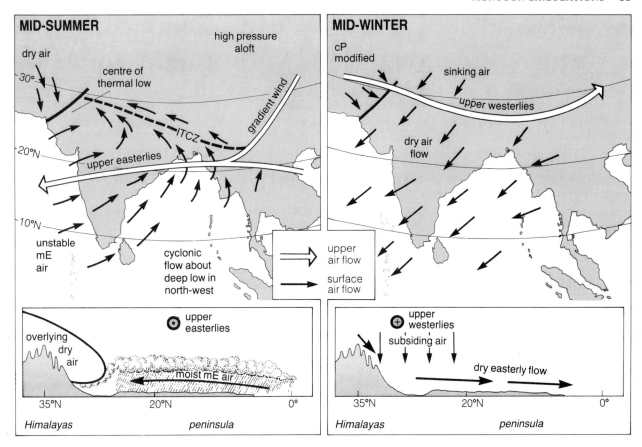

Fig. 4.27 The remarkable seasonal wind reversals in south-east Asia and the influences of the strong upper air flows.

above sea level and becomes a high altitude heat source. High pressure becomes established in the upper air immediately above. Thus a north-south gradient exists in the upper air. As this air flows southward it is deflected towards the west, so that an easterly jet stream is developed at a height of some 15 km.

The easterly jet stream flows across south-east Asia, above central India and on over the southern part of the Sahara. Convergences over Saudi Arabia and north-east Africa lead to subsidence and help to emphasise the aridity. Over India the swing of this upper air stream causes air to rise on the northern side and descend to the south, which partly accounts for the much drier conditions in peninsular India. Small waves travel through the general flow of the easterlies and also make for changeable weather during the wet monsoon period.

Exceptionally heavy rain occurs where unstable mT air meets relief barriers (p.59). Convection above heated plateaus and plains also adds to the intensity of precipitation.

The total rainfall received decreases markedly from the Bay of Bengal to the north-western plains. In fact the hottest part of the sub-continent has a low rainfall. There, subsiding air advances aloft from the north-west and overlies the

shallower moist monsoon air. Its blanketing effect prevents excessive vertical movements of the moist air.

The rainy period lasts in most places until mid-September, though during the 'rains' there are breaks, which vary in length from place to place and year to year. Gradually, subsiding air re-establishes itself over northern India, and the low pressure zone of convergence retreats southward. But during the retreat of the monsoon tropical cyclones are apt to develop within the low pressure zone, and violent storms may affect southern India during October and November. Storms are also likely during April and May, before the break of the monsoon.

The monsoon of eastern Asia

During summer humid air moves in, mainly from the east and south-east. But in winter strong, persistent flows of cold, stable air reach eastern Asia from the west and north-west.

The summer monsoon is weak compared with that of southern Asia. In the extreme south-east hot, humid mE arrives from the southern ocean. South-west China also receives moist air from part of the Indian monsoon system; though by the time it reaches China it has lost much of its moisture content. It brings short periods of torrential rain and occasional thunderstorms.

Over much of eastern Asia the rain is frequently

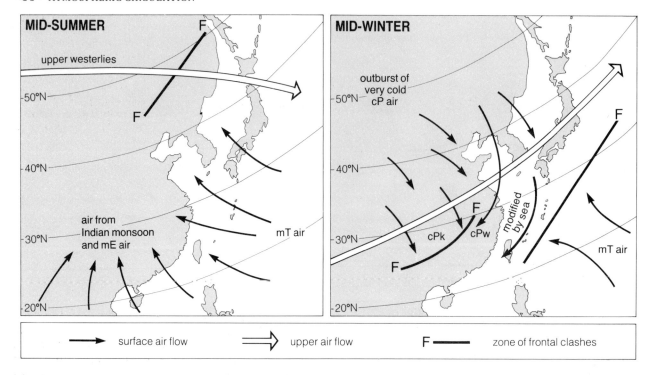

| → surface air flow | ⇒ upper air flow | F —— zone of frontal clashes |

Fig. 4.28   The east Asian monsoonal changes show a strong contrast between the cold continental outflow in winter and the humid oceanic inflow in summer; though north of latitude 40° moist air only occasionally penetrates far inland during summer.

of frontal origin. Weak disturbances cause much of it, and typhoons bring about a third of the summer rain to the coastal south-east, where prominent relief also makes for heavy rainfall.

In the northern parts of this huge area the inward flows are fairly shallow. Rainfall occurs close to the mean position of the polar front, where mT air of Pacific origin meets mP air from the Sea of Okhotsk. In the heart of the continent dry air moves eastward; but occasional surges of moist air provides some of the meagre rain which falls in the Asiatic interior.

During winter cold, dry cP air arrives in outbursts from central Asia. This is stable air: the skies are clear, except for wind-borne dust particles, and the weather bitterly cold. Some of this air swings clockwise around the general anticyclonic system, and reaches China with moisture gained from the western Pacific. Its origin is polar continental, but it becomes sufficiently unstable to produce cloud and precipitation over the coastal regions, notably where relief promotes uplift.

In central and southern China clashes between this mP air and cP air coming direct from the interior create disturbances along the fronts of contact.

At higher levels westerly winds prevail over China as a whole. Surface air from the high Tibetan plateau moves eastwards through the year, warming on descent and giving bright sunny weather in Yunnan.

4.17   Seasonal wind changes in West Africa

These wind changes can best be seen in terms of a northward migration of the ITCZ. **During summer** air from the Gulf of Guinea brings cloudy, muggy conditions over the coast. Storms develop as the moist air surges northwards.

There is little rain-giving disturbance where the air meets the dry subsiding cT air from the desert, as this largely overrides it, which makes for stability. There the wedge of moist air is of insufficient depth to produce many showers. But within the deeper mass of moist air to the south, belts of storms with towering cumulonimbus and squally gusts move westward, bringing heavy rain.

The upper tropical easterlies, originating in south-east Asia, pass high above the moist inflowing air. There is also another easterly flow in mid-troposphere, somewhat north of this. This is related to temperature differences; for a strong thermal gradient develops between the very hot cT air and the muggy mE air from the south. These upper air flows create wave disturbances with alternating convergence and divergence aloft, and the latter allow upward surges of moist air, bringing heavy rain.

In winter the Saharan cT air extends almost to the coast. The dry dusty north-east Trade winds (the Harmattan) blow from the north, with a dessicating effect over all but coastal regions.

The extent to which the rain-bringing disturbances can invade the semi-arid Sahel zone is of annual concern to its people, who depend on summer rain to provide pastures and establish crops. The erratic occurrences and the results are discussed further on p.152.

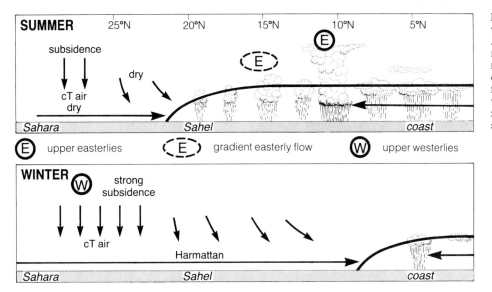

Fig. 4.29 While dry, dusty winds dominate most of West Africa during the winter months, in summer heavy rain falls in the low latitudes, decreasing northward in frequency and duration. The vulnerability of the Sahel zone to drought is emphasised in Figs 11.11 and 11.12.

4.18   Tropical east-west circulations

Besides the thermally-driven north-south Hadley cells, there are east-west circulations within the tropics. They cover broad areas, associated with strong convectional activity. Convectional uplift involves large-scale low-level inflows of air, with outflows in the upper air. One important convective source is the south-east Asian monsoon. This feeds a high-level west-east circulation, which involves subsidence over the tropical eastern Pacific and a returning westward air flow near the surface. Fig. 4.30 shows that, in the months that follow, there are other successive, increasingly strong, convective inputs from specific areas.

Fig. 4.30 **(A)** Air from the Asian monsoon flows eastward at high levels. Cyclonic storms, developed in unstable air over the western Pacific, increase in frequency during the southern summer months, and air spirals up to join the eastward flow. **(B)** A simplified, diagrammatic, view of circulations in a 'Walker Cell' in the southern hemisphere.

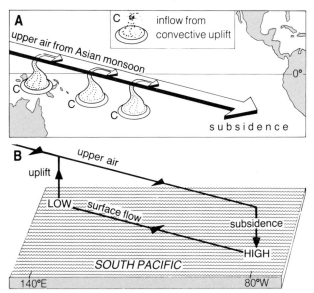

A strong Asian monsoon seems to intensify subsidence in the eastern parts of the South Pacific. In some years, however, when there is a weaker Asian monsoon, there is strong convection over the tropical mid-Pacific but weaker subsidence, and so lower atmospheric pressure, over the eastern equatorial Pacific.

A large-scale east-west tropical circulation in the South Pacific was recognised by Gilbert Walker in 1923. Later observers have named this the **Walker circulation**. Today, remote sensing by satellites confirms links between sea-surface temperatures and precipitation fluctuations in these pressure cells over the Indian Ocean and eastern Pacific.[1]

Variations in tropical convective inputs are closely linked to phenomena in other parts of the global climatic system. Lower than usual sea-level pressures and warmer sea-surface temperatures in the tropical eastern Pacific seem to coincide with higher pressures and prolonged droughts over central Australia, and parts of southern Africa and eastern Brazil.

Events outside the tropics, such as changes in the Rossby wave zonal index and storm track positions in mid-latitude, have links with anomalous weather within the tropics. Low-level incursions of cold air from Asia seem to increase convection during the south-east Asian monsoon; and cold surges from the southern Pacific appear to strengthen activitity in the Australian monsoon. In each case activity in the Hadley cell circulation is intensified, transferring greater amounts of heat energy from the tropics at high level.

Satellite-computer innovations are helping to give a clearer picture of global climatic relationships. **Circulation models** of the atmosphere are used to simulate climatic processes in detail. By simulating unusual surface temperatures in specific ocean areas, it is possible to observe probable responses in pressures and precipitation in distant parts of the globe. Such methods are being used in connection with the sudden anomalous appearance of the warm El Niño current off coastal Peru, and seem likely to allow successful predictions.

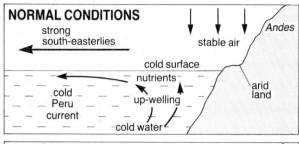

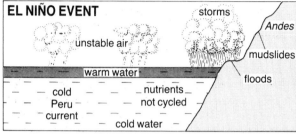

Fig. 4.31   The warm event in the eastern part of the South Pacific causes both physical dangers to the Andean coastlands and disruption to the oceanic nutrient cycle, which affects the huge fish population of the offshore Peru current.

4.19   East-west circulations and the El Niño current

The northward flow of cold water off the coast of western South America, the Peru current, is part of the normal anticlockwise ocean water circulation in the South Pacific. Off northern Chile and Peru persistent offshore winds drag the surface water westward, allowing colder water to well-up from below. As a result, sea-surface temperatures are some 5 C° cooler than in the western Pacific.

In this area of general air subsidence, the cold sea increases stability. The coastlands are arid, even though at times mist and low-level cloud, from condensation over the cold water, move inland. In general, however, air chilled by the cold surface continuously moves westward, as part of the south-east Trade wind, gaining heat and moisture as it flows across the South Pacific. Eventually it rises in the western arm of the Walker circulation.

The surface waters are normally moved by the wind, as a shallow oceanic flow, towards Indonesia. But every few years the Walker circulation becomes so weak that this surface flow is reversed. Warm surface water from the west then overrides the colder waters off southern Ecuador and Peru. This **'Warm Event'** occurs about Christmas time, so the warm current is known as **El Niño**, the Christ child. Its effects on these coastlands have often been disastrous, particularly those of the persistent El Niño of 1982–83, which led to widespread devastation.

Humid, unstable air, warmed by surface water moving from the western Pacific, brings torrential rain to the coastlands and Andean slopes of Ecuador and Peru. High snows melt and add to the

Fig. 4.32   The major atmospheric circulations are interlinked. The effects of a weak Asian monsoon and the related warm event in the South Pacific are usually accompanied by prolonged drought in these other countries.

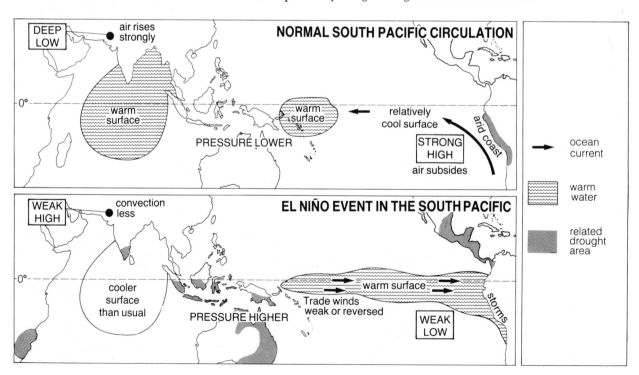

torrents rushing down erosion channels. Dry boulder-strewn river courses receive streams choked with material from loose, dry screes of the mountain slopes. Mudflows and flood waters overwhelm coastal settlements, destroying crops and sweeping away bridges.

In normal times the cold offshore waters are rich in nutrients carried towards the surface by deep, cold upwelling water. Plankton abounds and supports enormous numbers of fish and fish-eating birds. In the 1970s coastal fisheries were taking some 180 000 tonnes of fish *per day*. Even these rich resources were being over-fished. The El Niño breaks the food cycle, for the nutrients are no longer cycled upward. The 1982–83 Warm Event proved disastrous to an industry hit by over-fishing.

Also during the 1982–83 reversal of the Walker circulation there was prolonged drought in Australia, with livestock starving in pastoral areas; dust storms from the interior even affected the Victorian coastland. In southern Africa drought affected the wildlife, domestic cattle and crops, and there was severe malnutrition in tribal homelands.

At the same time unusually strong hurricanes struck the western islands of the South Pacific. Great updraughts of air enhanced the Hadley circulation. In the long run, this would tend to cause El Niño to decline: for a strong north-south circulation strengthens the Trade winds, so that surface water begins to flow westward once more, re-establishing colder water in the eastern Pacific.

4.20 Ice caps and global circulation

Both the Arctic and Antarctic play a large part in maintaining the global heat balance. Each gains energy only during the summer period and radiates heat throughout the year. There are long periods of continuous insolation, though the oblique rays give a low intensity heat. Snow and ice surfaces reflect back about four-fifths of the solar energy received. In clear weather, radiation losses through the dry air are particularly great. Yet each year the outflow of energy must be nearly balanced by an inflow, or the polar regions would become progressively colder.

Much of the extra energy arrives by advection, that is by horizontal movements of warm air and water vapour, which help to balance the loss by radiation. Vast amounts of latent heat are released as water vapour condenses to droplets and the drops solidify. Over Antarctica the average precipitation, mostly in the form of small, hard snow particles, is estimated as the equivalent of 150–250 mm of water, though only about 50 mm is recorded near the South Pole.

Antarctica is a landmass with bordering oceans, and covers about three times the area of the Arctic. There are few incursions of warm air into the Antarctic in winter. The strong westerlies in the lower atmosphere form a zonal belt in the higher middle latitudes. Also the continents do not extend from tropical to polar latitudes in the southern hemisphere. However, eastward-travelling cyclonic storms do move in from time to time, causing heavier snowfall, and small depressions form around the continental edge, though few move over the central region.

In winter there is a low pressure vortex in the upper air, far above the cold dense air masses over the Antarctic plateau and the South Pole, with high pressure over the perimeter of the continent. Again, strong zonal winds do not readily allow an exchange of air with lower latitudes.

ANTARCTIC WINTER

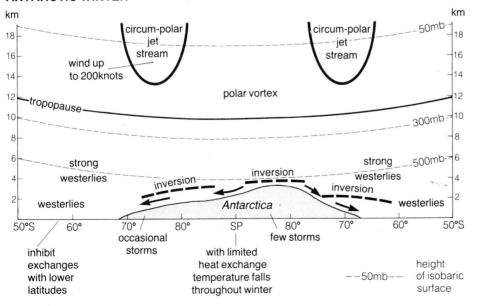

Fig. 4.33 The strong zonal flow of the surface westerlies tends to isolate the continent from the influx of heat energy. Great swings of the upper westerlies do occasionally allow depressions to cross parts of Antarctica, which introduce relatively warm air and increase the snowfall.

The earth is nearest the sun during the Antarctic summer, so that about 7 per cent more solar energy is received by the Antarctic atmosphere in summer than by the Arctic atmosphere during the northern summer. Also the Antarctic surface is at such an altitude that solar radiation has to pass through only about two-thirds of the mass of air which covers the northern polar surface. This has some relevance to the concern felt over the discovery of a large hole in the protective ozone layer above Antarctica, considered on p.150; for increased melting of the polar ice-caps can have worldwide repercussions.

During summer the low in the upper air is replaced by a high, and at low levels occasional large cyclonic storms are able to bring heat and moisture into the continent.

The surface temperatures in the Antarctic vary with latitude and altitude. Mean annual temperatures at the Soviet base at Vostok, which stands 3300 m above sea-level, are −56°C, compared with −51°C at the pole, a thousand kilometres further south, but 600 m lower. For three successive years Vostock recorded minimum temperatures between −86°C and −88°C.

In winter inversion occurs in the air above the cold surfaces, and strong down-slope winds develop, usually up to a hundred metres or so in depth. When they reach a certain velocity gustiness sets in, causing intense local blizzards.

The Arctic climate is more variable, for it is an ocean region with only a land perimeter. The Arctic air holds more water vapour, and so the loss of long-wave radiation tends to be less than in Antarctica. Heat from the underlying ocean passes through the ice and helps to warm the atmosphere. Also in summer the land surfaces about the Arctic contribute more energy to the atmosphere than the water surfaces surrounding the Antarctic landmass.

Mean annual temperatures on the Greenland ice-cap are of the order of −30°C to −35°C, but about −20°C to −25°C over the surface of the Arctic Ocean. However, with land and sea areas alternating, there are considerable climatic differences within the Arctic. The zonal westerlies are interrupted from time to time as cold air surges southward. Clashes between mP and cP air occur, causing frontal disturbances and precipitation, especially during summer.

4.21   Oceans and atmospheric energy

Oceans act as heat reservoirs. Their importance in transferring energy to the overlying atmosphere has been stressed by reference to both El Niño and the Arctic climate. In the higher and middle latitudes especially, warm currents yield vast amounts of heat, both directly and as potential energy in moisture, to the air above. In general, wind drag on surface waters helps to create and maintain an oceanic circulation and the poleward transfers of heat. As we have noted, density differences due to heating of tropical surface waters and cooling of polar waters also give rise to slower circulations, at the surface and at depth.

In the outer tropics the strong inputs of solar energy combine with steady prevailing winds to cause rapid evaporation, so increasing the salinity of the ocean's surface. Nearer the equator, however, moister air with potential instability, and greater cloud and rainfall, make for a lower salinity.

The El Niño temperature anomalies emphasize how energy transfers between the air and offshore waters affect coastal climates. Fog banks and low cloud are not only a normal condition off northern Chile and Peru, but also along the cold water coasts of California and south-west Africa. The summer fogs of San Francisco are notorious. Cold, dense, fog-laden air pours in through the Golden Gate at such low levels that the hill summits to the north, still in bright sunshine, have an average summer temperature 5 C° above that of the city

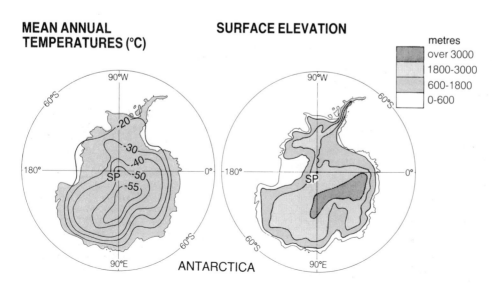

MEAN ANNUAL TEMPERATURES (°C)

SURFACE ELEVATION

metres
over 3000
1800-3000
600-1800
0-600

ANTARCTICA

Fig. 4.34 The effects of relief, and of isolation from low-level inflows, are seen in the mean annual air temperatures over the Antarctic continent.

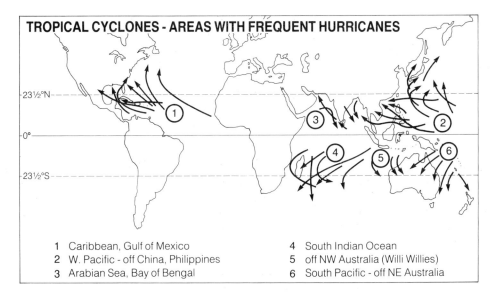

TROPICAL CYCLONES - AREAS WITH FREQUENT HURRICANES

1. Caribbean, Gulf of Mexico
2. W. Pacific - off China, Philippines
3. Arabian Sea, Bay of Bengal
4. South Indian Ocean
5. off NW Australia (Willi Willies)
6. South Pacific - off NE Australia

Fig. 4.35 Intense cyclonic storms occur mostly over the western parts of the oceans, towards the outer tropics. Nearer the equator the Coriolis force is weak; so few storms of this type develop in the low latitudes – though distant cyclones may create tidal waves which travel hundreds of kilometres, and are of sufficient amplitude to swamp low-lying atolls.

below. The strong advective flow through the break in the coastal hills is influenced by thermal low pressure over the hot surfaces of the Great Valley to the east. Inland the fog clears rapidly.

4.22 Oceans and tropical cyclones (typhoons)

Tropical cyclones are systems with almost symmetrical swirls of strong hurricane-force winds about a calm centre. They have no marked fronts between cold and warm air. They form over oceans in the lower middle tropics, but seldom develop within seven degrees of the equator, where the Coriolis force is slight. As their centres move from east to west, the whole system may extend to 50–1000 km across. They are most frequent in the eastern parts of the oceans in late summer and early autumn. In the Indian Ocean most occur just before and just after the wet monsoon season.

Usually they only form over water with a surface temperature above 27°C – a source of heat and moisture. Troughs of low pressure move westward across the tropical Pacific and Atlantic, often with clusters of convective cells and cumulus clouds with strong vertical development. Some are triggered into forming circular vortices, perhaps where upper air divergence favours upward movements of surface air. This is increasingly likely away from the upper air subsidence associated with the Trade wind source regions.

Energy released by condensation helps to create spiral bands of towering cumulonimbus, with heavy rain, about the storm centre. Low frictional drag at the sea surface means that there is little to impel air to flow across the isobars towards the centre. Heat is given out in the huge cloud banks about the clear 'eye' of the storm, and acquired by air which descends, warming adiabatically, in the calm centre.

SECTION

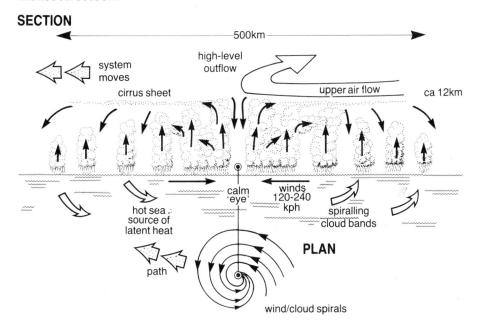

Fig. 4.36 As the whole system moves westward, the individual cloud bands spiral inward, as shown, about the eye of the cyclone. In a deepening system, the air ascending is removed by upper air currents some 12 km above the surface. As long as the system passes over a very warm water surface, condensation, releasing latent heat energy, can fuel it and keep it active.

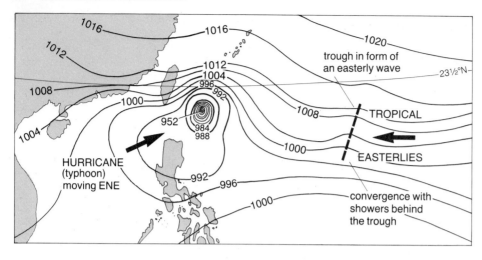

Fig. 4.37 A typhoon moving eastward towards Taiwan, with a following wave disturbance in the Trade wind flow over the western Pacific.

Other convective cloud masses giving heavy rain, also spiral about the centre, with air descending on their flanks. The rainfall of the whole system varies considerably, but is torrential in character. As much as 250 mm may fall during the passage of a cyclone, sometimes five times that amount.

As the ascending air reaches the tropopause, it spreads outward, forming a wide layer of cirrus cloud above the storm system. The pressure gradients about the centre sustain winds of great force, with 180° change of direction as the eye passes over a location. Tropical hurricanes frequently cause great structural damage and coastal flooding. The winds generate tidal waves, and the central low pressure, of the order, perhaps, of 880 mb, can uplift water into a wave surge.

In the global context, hurricanes play a part in transferring energy poleward from the tropics.

Tornadoes are much smaller systems, some only 100 metres across; but they, too, can be violent and extremely destructive. They are air-spirals of tremendous velocity, with vigorous up-draughts, and form over hot surfaces in the outer tropics and near tropical lands, mostly when the air is moist and unstable, especially during hot afternoons.

Tornadoes writhe their way across country at some 50 kph, detaching and sucking up loose debris. Frequently a funnel-shaped cloud develops above them, opening out to a heavy cumulonimbus mass above. Thunderstorms, with heavy rain and large hailstones accompany them when the air is moist.

They occur in other parts of the world, even occasionally in Britain. In southern USA, especially in the Mississippi valley, fierce tornadoes have devastated small townships. When they move over the sea a **water spout** may develop, carrying water up towards the overhanging cloud, and flinging it out again by centrifugal force, frequently with fish as well.

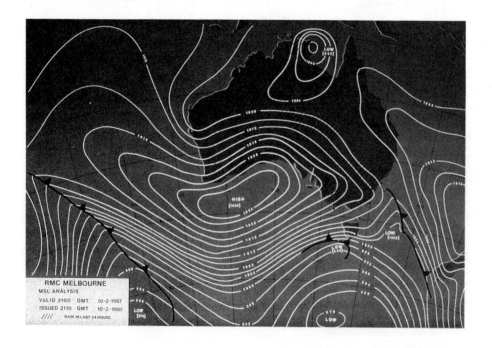

Fig. 4.38 A meteorological map for 10 February 1987, issued by RMC Melbourne, shows conditions extending from the cyclonic disturbances in the low latitudes to the depressions developed far to the south in the temperate westerlies zone.

4.23 Cyclones and the zonal weather patterns

The cyclonic disturbances described above can now be seen in the context of global circulations. Figs 4.38 and 4.39 show mid-summer meteorological conditions about the Australian landmass. High pressure is centred over the Gulf of Carpentaria, so that the coasts of Victoria and New South Wales are cloudy, with onshore winds. Further south, frontal disturbances in the zone of the temperate westerlies are directly affecting Tasmania and South Island, New Zealand.

By contrast, storms are well developed across the low latitudes, over Indonesia and the Pacific, and a **tropical cyclone** with spiralling cloud bands is moving over Arnhem Land. Bright areas on the satellite picture show convective cloud formations over Western Australia, and Fig. 4.38 indicates those parts of the arid interior which have re-

ceived welcome rain showers over the past 24 hours.

The cloudless, dry conditions affecting most of southern Australia will gradually move eastward over Victoria and New South Wales.

This particular season brought severe cyclones to northern and western parts of Australia. Fig. 4.40 shows the passage of cyclone Elsie,

Fig. 4.39 A geostationary satellite view of the cyclonic storm seen over Northern Territory in Fig. 4.38. Other storms have developed north of Australia in the hot moist air over Indonesia and the western Pacific. An upper air high above the Indian Ocean is causing storms to move over the coast of Western Australia – as happened to cyclone Elsie a fortnight later (Fig. 4.40). A great swath of cloud affects central Australia. High pressure covers the south of the continent, though Tasmania is in the path of a depression in the westerlies, and a front, bringing cloud to New South Wales, approaches New Zealand.

which developed ten days after the situation shown in Fig. 4.38. It moved westward, deepening, until diverted southward by an upper level high over Northern Territory. As it swung eastward over the coast, a barograph at the badly damaged Mandoora homestead recorded the trace shown in Fig. 4.41. Winds gusted to 222 kph; a speed which, according to records, has only been exceeded by a gust of 248 kph when cyclone Trixie devastated parts of Darwin in 1975.

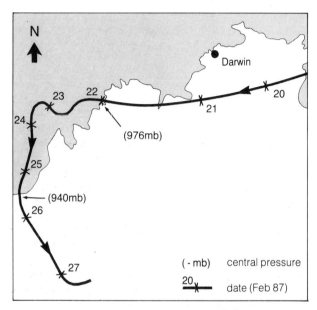

Fig. 4.40   The passage of cyclone Elsie in February 1987. It developed over Northern Territory and, continuing to deepen, moved onto the coast of Western Australia. It destroyed an isolated homestead, which, nevertheless, recorded conditions at the 'eye'. Hundreds of horses and cattle were lost, but fortunately no major settlements lay in its path.

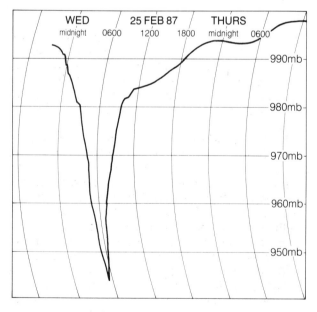

Fig. 4.41   The barograph trace of cyclone Elsie at Mandoora homestead.

5

WORLD CLIMATES

5.1 Climatic regions

There are, of course, innumerable local climates. Yet, on a global basis, areas far removed from each other have such striking similarities in their climatic elements and regimes that we regard them as similar climatic regions.

Rarely, if ever, do all the climatic elements in one location have the same characteristics as those in another. In order to classify climatic regions, we need to identify significant common climatic features. But, of course, similar mean figures for rainfall or temperature over a given period can be misleading. In one area persistently high humidity, cloudiness, and steady rain can give the same monthly rainfall figure as that of an area where heavy storms alternate with dry spells. The former may have small daily and monthly temperature ranges, the latter considerably larger ones, yet share the same mean monthly temperature. In time, their particular characteristics will be reflected in geomorphological features, vegetation and human responses; so that, despite similar mean figures, and even similar macro-regimes, the regions are noticeably dissimilar in many respects.

Defining climatic regions by whatever criteria is bound to be hazardous and run the risk of oversimplification. There is seldom a sharp climatic divide in nature. Transitional zones are usually found between one recognisable climatic region and the next. But while sub-division might make for greater accuracy of description, there is always a practical limit to the number of sub-divisions: no two places have exactly the same climate.

5.2 Finding a basis for classification

Regional climatic characteristics are apt to be reflected by the vegetation. So numerical values for the temperature and precipitation which will allow particular plants to flourish may be chosen to define the limits for climatic types. The combination of precipitation and temperature is usually considered on a seasonal basis; for a certain quantity of rain received during a hot season is likely to be less effective for plant growth than the same amount received during a cooler season, when the rate of evaporation is less.

Some classifications use values of **precipitation effectiveness**, obtained by determining ratios of precipitation to temperature. Others consider the *actual* values for evaporation over a given period. Accurate knowledge of the loss of water from moist soil and plant surfaces would be particularly valuable for classification purposes if measured regularly at widespread stations. But unfortunately measured evaporation figures are not available for many parts of the world.

Of course the composition of the vegetation is not necessarily a simple response to climate. Infertile soils, poor drainage, and interference by people, animals, and fire are among the complicating factors.

5.3 Köppen's classification

In 1918 Dr W. Köppen published a classification, which he later modified from time to time. In this he considered **precipitation effectiveness** for plant growth as a major factor. So he chose appropriate mean values of both temperature and precipitation to define the limits of his climatic groupings.

Köppen observed the conditions of growth required by various groups of plants, ranging from the **megatherms** which favour warm habitats, through the intermediate **mesotherms**, to **microtherms** which thrive in a colder environment. By relating them to various temperature limits, he defined certain main climatic groups.

He distinguished arid and semi-arid lands on the basis of evaporation exceeding precipitation, with sub-groups categorised on a temperature basis. The polar and near-polar regions, with long winters and short summers, and mostly beyond the limits of tree growth, he also defined by a limiting mean temperature figure.

The letters *A* to *E* were used to designate five main groups, which he described as follows:

A	Tropical Rainy:	a hot climate with no cool season; the average temperature of each month is over 18°C
B	Dry:	evaporation exceeds precipitation
C	Humid Mesothermal:	the warmest month has a mean temperature above 10°C; the coldest month has a mean temperature between −3°C and 18°C
D	Humid Microthermal:	the warmest month has a mean temperature above 10°C; the coldest month has a mean temperature below −3°C
E	Polar:	no month averages over 10°C

The *D* type, for instance, recognises that a 10°C average for the warmest month coincides *roughly* with the poleward limits of forest; and that the −3°C mean for the coldest month is near the equatorward limit of frozen ground, where snow remains on the surface for at least a month.

Sub-groups

A second, small letter is used to describe the rainfall distribution. In *A* climatic regions, *f* indicates that no month has a mean rainfall of less than 60 mm; *w* shows that at least one month has under this amount. The values for the *Af* tropical rainy climate are those which Köppen believed would support tall tropical rainforest. Other small letters are used, such as *m* for a monsoon climate which supports rainforest despite a short dry season.

The *B* climates (dry ones) are also sub-divided into a *BW*, *arid*, type (W = Wüste, desert) and a *BS*, *semi-arid*, or steppe, type. The *BS/BW* limits are identified from formulae, using different constants combined with rainfall and temperature values for each sub-type.

Other small letters are used as third symbols to aid description: thus *BWh* and *BWk* (*h* = heiss, *k* = kalt) represent deserts where mean annual temperatures are over and under 18°C respectively.

Various minor characteristics of the *C*, *D*, and *E* climates are also shown by additional symbols. G.T. Trewartha's modification of Köppen's classification within the framework of the five major groups, shows the use of symbols in more detail (pages 56–8).

Limitations

It must be stressed that Köppen's form of empirical classification makes no attempt to take account of the *causes* of the climate described, nor of the relations between the location of the climatic region and those of air mass source regions, or other likely influences. Trewartha stresses the value of supplementing empirical classifications with some description of the mode of origin of the climatic features used to define the boundaries.

5.4   Thornthwaite's classification

Precipitation effectiveness

During the early 1930s, C.W. Thornthwaite put forward, and modified, a system of classification which also looks on plants as 'meteorological instruments'. He derived a measure of **precipitation effectiveness** – the P/E Index – by dividing the total monthly precipitation (P) by the total monthly evaporation (E), and then adding the twelve values for each month.

The main drawback, again, is that the actual evaporation data are not available for many parts of the world. As a substitute, Thornthwaite used readings of temperature and precipitation from various stations in south-west USA and derived a formula which served to give the P/E Index. This formula was then used for other parts of the world.

From the calculated P/E values he distinguished five **humidity provinces**, each with characteristic vegetation:

Humidity province		Vegetation	P/E Index
A	wet	rainforest	over 128
B	humid	forest	64–127
C	sub-humid	grassland	32–63
D	semi-arid	steppe	16–31
E	arid	desert	under 16

Small letters indicate the seasonal **concentration of precipitation**: r = abundant in all seasons; s = deficient in summer; w = deficient in winter; d = deficient in all seasons.

Thermal Efficiency

Thornthwaite also developed a formula for calculating **thermal efficiency** – the T/E Index – and, with values again grouped from 0–128, he recognised six **temperature provinces**:

A′	tropical	D′	taiga
B′	mesothermal	E′	tundra
C′	microthermal	F′	frost

Combining the effects of precipitation, evaporation and temperature

The climate is described by presenting Thornthwaite's groups together. For his world map of climatic regions, Thornthwaite chose 32 of the 120 possible combinations, as in Fig. 5.1.

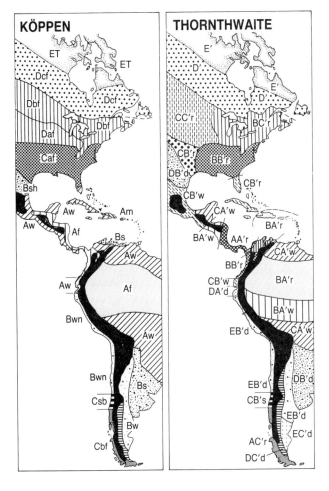

Fig. 5.1 American climatic regions defined on the basis of Köppen's classification, broadly based on temperature/precipitation/evaporation conditions associated with various groups of plants; and also on Thornthwaite's ideas of effectiveness of precipitation for plant growth, depending on air temperature and evapo-transpiration.

Climatic classification helps to define particular regions and allows us to compare and contrast their conditions with those of others; but, as you can see, much depends on the basis used for classification. There are no absolute regional boundaries; but, nevertheless, regional patterns such as these help us to concentrate our attention on the climatic characteristics of particular parts of the earth's surface.

Modern approaches to climatology allow us to examine the processes working in climatic systems, and to predict changes in the long- and short-terms. The descriptive approach, of climates based on mean climatic conditions, are especially valuable when the regions defined are seen to include dynamic systems of a particular kind.

5.5 Evapo-transpiration

In 1948 Thornthwaite put forward a classification which considers the losses of water from soil and plant cover as evapo-transpiration. He used a formula to find the loss by evaporation and transpiration which would occur if water were *always* available to the roots. In fact the loss was calculated as a function of the temperature.

This **potential evapo-transpiration** can then be compared with the precipitation and climatic boundaries defined on that basis. Again the drawback is the difficulty of obtaining values for the loss of moisture to the atmosphere: hence the need for the calculation in terms of temperature. The chief use of the potential evapo-transpiration values has been for single stations, for which curves drawn for monthly values can be compared with those for precipitation, so that periods of water surplus or water deficiency can be gauged.

There are thus means of estimating the evapo-transpiration – the loss of water from soil and plants – and instruments *can* also be used to measure the **potential loss** from the surface of short, green, growing vegetation. This is done by sinking a gauge in a reasonably level, open, grassed plot, in such a way that it becomes a typical part of the surface.

This **evapo-transpirometer**, or **lysimeter**, consists of two watertight tanks filled with soil above a layer of gravel, and sunk into the ground. Each surface is sown with grass like the surroundings, or with similar short green vegetation. The base of each tank is connected to a collecting can housed in a third watertight tank. Water can only enter the upper soil-tanks from the atmosphere, and can leave only by the outlets at the bottom.

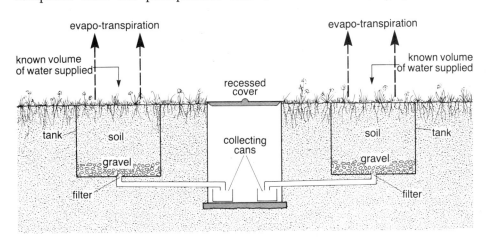

Fig. 5.2 The lysimeter measures the moisture supplied to maintain a complete local plant cover against its losses by evapo-transpiration.

Each day the tanks receive a known amount of water from precipitation, if any, (measured by a rain gauge) and from sprinkling, used initially to saturate the soil in the tanks and then to maintain sufficient soil moisture to satisfy plant needs. The amount that percolates into the collecting cans is measured, and can be converted into depth in millimetres for each soil tank. The difference between the amount entering and the amount leaving the tanks represents that lost by evapo-transpiration. A complete plant cover is maintained.

5.6   The soil-water budget

Fig. 5.3 shows how in a humid, cold winter/warm summer, mid-latitude location the quantity of soil-water changes through the year in relation to the water-holding capacity of the soil.

The quantity of water held in the soil-water zone, from which plants can draw moisture, is the **soil-water storage**. This is decreased by evapo-transpiration (−G) and recharged (+G) by precipitation (P). As the soil becomes saturated, any surplus (R) may move downwards to the ground-water zone or flow as surface run-off.

The *actual* evapo-transpiration (Ea) is that returned to the atmosphere from the ground vegetation, which adapts to avoid excessive water loss. The *potential* evapo-transpiration (Ep) is the water needed to maintain the storage capacity of the soil and its uniform cover of green vegetation. This could be supplied by precipitation or by irrigation,

so the difference between the potential and actual quantities (Ep−Ea) is the *extra* that would have to be supplied **to maintain maximum plant growth**, or the **soil-water deficit** (D).

The air temperature and humidity affect the actual losses by evaporation. At a time of soil-water deficit, plants may die down or become dormant, and so reduce the amount lost through evapo-transpiration.

Fig. 5.3 is a simplified model for a northern hemisphere mid-latitude location, with cold winters and warm summers. It assumes that there is uniform precipitation through the year. The potential evapo-transpiration regularly responds to the changing seasons: in mid-summer much water is needed to make up the losses, and in mid-winter the minimum amount is needed.

From mid-December to April there is surplus water. Then plants begin to withdraw water from storage, for Ea exceeds P. In a short while the amount the actual vegetation withdraws is less than that which would be lost from a maximum plant cover. So **to ensure maximum growth** the soil-water deficit (D=Ep−Ea) would have to be made up by irrigation.

During the autumn P again exceeds Ep, so there is a period when the soil-water storage is recharged. When its storage capacity is reached, the period of winter surplus begins once again.

Calculations and expressions of this kind are obviously most useful for assessing the benefits to be gained by increasing irrigation. Relatively few parts of the world have sufficient precipitation to provide the soil-water needed for maximum plant production throughout the growing season.

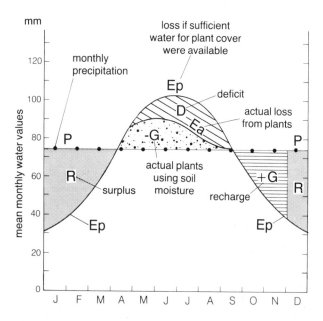

Fig. 5.3   The soil water budget, expressed in terms of the actual evapo-transpiration losses (Ea) and the quantity required to maintain the maximum plant cover (Ep).

5.7   Trewartha's modification of the Köppen system

Group *A*
TROPICAL RAINY CLIMATES: temperature of the coolest month over 18°C

Af   *no dry season*: driest month has over 60 mm; ITCZ with mT or mE air movements

Am   *short dry season*: but rainfall sufficient to support rainforest; (wet monsoon type)

Aw   *dry during the period of low sun (winter)* – driest month under 60 mm: dry cT air in winter; wet during period of high sun when ITCZ moves poleward and moist mT air flows in

Group *B*
DRY CLIMATES: evaporation exceeds precipitation (*W* – arid; *S* – semi-arid)

boundary between *BW/BS* is where, for winter precipitation maximum, r/t = 1; for summer maximum, r/t + 14 = 1; for evenly distributed precipitation, r/t + 7 = 1.
(r = annual rainfall in cm; t = mean annual temperature in °C)

BWh   *hot desert* – mean annual temperature over 18°C; source region of cT air; dry cT winds

BSh   *semi-arid, tropical/sub-tropical:* cT air mainly, but short rainy season

BWk   *middle latitude interior desert:* cT air mass in summer, cP air mass in winter; large annual temperature range

BSk   *semi-arid middle latitude:* meagre rainfall, mostly in summer, when mainly cT air; cP air in winter (*n* – nebel – is used to show frequent fog along coastlands with cool water offshore)

Group *C*

HUMID MESOTHERMAL (moist temperate): coldest month between 18°C and 0°C (*f* – no dry season; *w* – dry winter)

Cs   *sub-tropical, dry summer:* at least three times as much rain in wettest winter month as in driest summer month; driest month less than 30 mm; summer dominated by sub-tropical high, cT air; winter with mP air, cyclonic storms and rain

Csa   *hot summer:* warmest month averages over 22°C

Csb   *warm summer:* warmest month averages under 22°C

TYPES OF CLIMATE

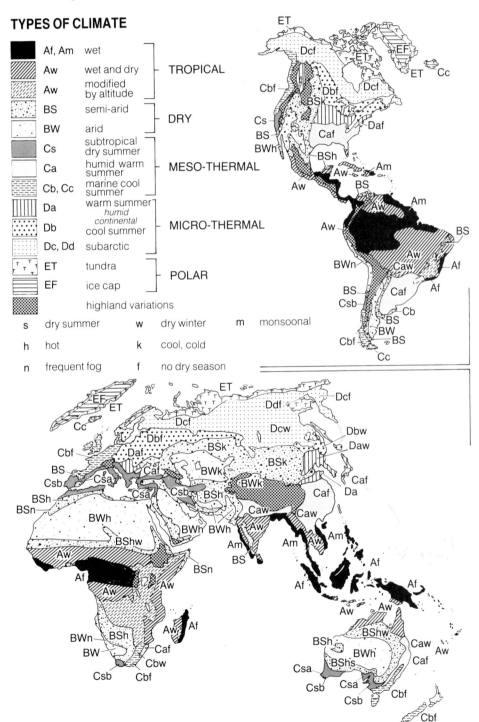

Fig. 5.4 The climates of the continents, based on Trewartha's modification of Köppen's classification. The system uses *H* for undifferentiated *highland climates*; stippling is added on maps to certain tropical/sub-tropical uplands of moderate elevation.

Ca   *humid sub-tropical, hot summer:* warmest month over 22°C; in summer moist, unstable mT air after ocean passage from sub-tropical high; in winter cP air invading, with cyclonic storms

Caf   *no dry season:* driest month over 30 mm

Caw   *dry winter:* at least ten times the rain in wettest summer month as in driest winter month

Cb   *marine climate, cool/warm summer:* warmest month under 22°C; mainly mid-latitude west coast; moist mP air; series of depressions; rain in all seasons

Cbf   *no dry season:* the most common *Cb* type; (*Cbw* describes parts of south-east Africa)

Cc   *marine climate, short, cool summer:* warmest month below 22°C; less than four months over 10°C; rain in all seasons

Group *D*
HUMID MICROTHERMAL (rainy/snowy, cold): coldest month under 0°C; warmest month over 10°C

Da   *humid continental, warm summer:* warmest month over 22°C; precipitation in all seasons; summer maximum; winter snow cover; frequent clashes between polar/tropical air; variable weather

Db   *humid continental, cool summer:* warmest month below 22°C; as for *Da*, long winter snow cover

Dc   *sub-Arctic:* warmest month below 22°C; less than four months over 10°C; winter cP air mass, cold stable air; summer occasional cyclonic storms with mP air; precipitation light; low winter evaporation, so remains moist

Dd   *sub-Arctic, very cold winter:* coldest month below −38°C; precipitation very light

Group *E*
POLAR: temperature of warmest month less than 10°C

ET   *tundra:* warmest month above 0°C; mP, cP, and A airmasses interact; cyclonic storms; light precipitation, mainly in summer

EF   *ice cap, perpetual frost:* no month over 0°C mean temperature; source regions for Arctic/Antarctic air masses.

CLIMATIC REGIMES

Climates of the low latitudes

5.8   Tropical rainy (*Af*)

This occurs in lowlands within 5°–10° of the equator, but extends to higher latitudes where the tropical easterlies, unstable after their passage over warm oceans, bring rain. By definition these climates support low latitude rainforest, though above 1000 m or so cooler highlands do not support a typical form of forest.

The noonday sun is never far from the zenith, and there is little variation in hours of daylight. However, high humidity and considerable cloud cover keep temperatures from soaring; the maximum is usually well below those of the dry, outer tropics during summer. The hottest periods occur during dry spells.

There is little seasonal variation: mean monthly temperatures are generally of the order of 26°–27°C, with a daily range of 8–10 C°. The monotonously high temperature combined with high humidity can be oppressive.

The mean annual rainfall is high. Over the oceans it contributes to the relative freshness of surface water in these latitudes, where the rate of evaporation is less than in the windier, sunnier, outer tropics.

The daily regime follows a regular pattern, especially during wetter spells. The somewhat cooler nights allow dew to form from air of high humidity, though skies remain clear. Morning haze, or low mist over swampy land, clears rapidly. By mid-morning cumulus builds up, and clouds increase in height and density until, during the heat

TROPICAL RAINY CLIMATES - Af and Am

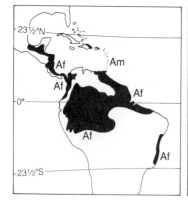
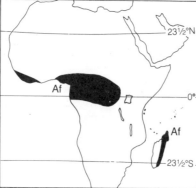
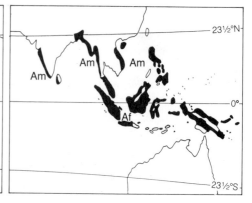

Table 5.1 Tropical rainy climates (*Af*).

Uaupes, Amazon lowlands. 0° 08′ S. 83 m													
	J	F	M	A	M	J	J	A	S	O	N	D	Total
°C	26	27	26	26	26	26	25	26	27	27	27	26	–
mm	262	196	254	269	305	234	223	183	132	175	183	264	2677

Max/min °C		Relative humidity %			
Year	Absolute	Hour	Year		
31	38	0530	97		
22	11	1230	74		

Singapore, South-east Asia. 1° 18 N. 10 m													
	J	F	M	A	M	J	J	A	S	O	N	D	Total
°C	26	27	27	27	28	27	27	27	27	27	27	27	–
mm	252	172	193	188	172	172	170	196	178	208	254	257	2413

Max/min °C		Relative humidity %			
Year	Absolute	Hour	Year		
31	36	0900	79		
23	19	1500	73		

Kisangani, Zaïre basin. 0° 26′ N. 417 m													
	J	F	M	A	M	J	J	A	S	O	N	D	Total
°C	26	26	26	26	26	25	24	24	24	25	24	25	–
mm	54	84	178	157	137	107	132	165	183	218	198	84	1705

Max/min °C		Relative humidity %			
Year	Absolute	Hour	Year		
30	36	0530	97		
21	16	1130	68		

of the afternoon updraughts of unstable air form cumulonimbus, with heavy thundery downpours. Later the skies clear once more.

Some days are cloudier than others, for slow-moving, shallow troughs and weak cyclonic systems sometimes develop in these latitudes of small pressure gradients, giving longer periods of rain. In some places there are considerable variations in rainfall amounts from year to year, and some months are noticeably drier, as in the eastern Amazon lowlands.

Parts of the east coasts of Brazil, central America and Madagascar are affected by moist, unstable, onshore air and have a somewhat similar climate. But the rainy periods are often more prolonged, and they experience more severe cyclonic disturbances and occasional hurricanes.

5.9 Tropical wet monsoon (*Am*)

These, with their dry seasons, have much in common with the *Aw* climates (p.60). Winter temperatures are usually very warm, but vary with location. Temperatures rise to a maximum just before the 'burst' of the monsoon, but fall somewhat during the cloudy, muggy, wet period, despite the higher altitude of the noonday sun.

Table 5.2 The period of very heavy rainfall during the summer months characterises these monsoon climates; but even during the dry winter maritime influences maintain a fairly high relative humidity, despite midday temperatures generally rising to more than 20°C. Compare this with the midday humidity during the dry season in the Aw climates (Table 5.3)

Bombay, Western India. 18° 54′ N. 11 m													
	J	F	M	A	M	J	J	A	S	O	N	D	Total
°C	23	23	26	28	30	29	27	27	27	28	27	26	–
mm	2	2	2	2	18	486	618	340	264	64	13	2	1808

Max/min °C			Relative humidity %		
Jan	May	Absolute	Hour	Jan	Jul
28	33	Jan Mar	0800	70	83
19	27	12 38	1600	61	83

Chittagong, Bangladesh. 22° 21′ N. 26 m													
	J	F	M	A	M	J	J	A	S	O	N	D	Total
°C	19	21	25	27	28	28	27	27	27	27	23	20	–
mm	5	26	64	150	264	533	597	518	320	181	56	15	2831

Max/min °C			Relative humidity %		
Jan	Jun	Absolute	Hour	Jan	Jun
26	31	Jan Mar	0800	81	87
13	25	7 39	1700	58	83

Fig. 5.5 (left) The classification includes the coastlands of Brazil and Madagascar, in the outer tropics, for they frequently experience onshore movements of unstable air, bringing periods of heavy rain throughout the year.

During summer, windward slopes receive an exceptionally large rainfall from inflows of mT air, while leeward areas are usually much drier. In a classification related to the precipitation effectiveness for plant growth, much of the Indian peninsula to the lee of the abrupt plateau edge of the Western Ghats has characteristics of an *Aw* or *Bsh* climate.

The Indonesian islands, Malaysia, and parts of Papua–New Guinea are sometimes seen as monsoonal sub-types of an *Af* climate, for marked seasonal changes in the direction of air flow give pronounced rain-shadow effects. Northern slopes have their lowest rainfall from May to September, when air flows northward into Asia. Southern slopes are well-watered during these months, but exceptionally dry from December to March, when air moves towards northern Australia.

5.10 Tropical climates with wet and dry seasons (*Aw*)

Poleward of the *Af* climates there is often a gradual transition to an *Aw* climate. The dry season becomes longer with increasing latitude, until conditions are semi-arid. Such a transition is most marked in the interior of landmasses, for warm ocean currents can extend humid coastlands polewards, and cold offshore currents can bring arid conditions into the lower latitudes, as in southern Ecuador.

During the cooler season the outer tropics are under the influence of cT air moving equatorward from the sub-tropical source regions. But as the midday sun approaches the zenith, temperatures

Fig. 5.6 Here in Indonesia abundant rain falling through the year on river-dissected highlands necessitates careful water control. With mean monthly temperatures above 25°C, skilled agricultural techniques are needed to regulate the abundant plant growth and to cultivate selectively a variety of tropical food plants.

Fig. 5.7 The simplified section reflects the changing location of the ITCZ, with convergence and incursions of less stable air during the hotter time of year, and more stable air becoming dominant during the less hot season.

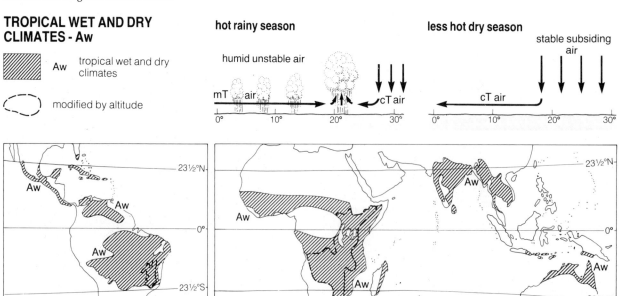

Sokoto, Northern Nigeria 13° 01′ N. 350 m													
	J	F	M	A	M	J	J	A	S	O	N	D	Total
°C	24	26	31	33	33	30	28	26	27	29	27	25	—
mm	—	—	2	10	43	94	152	244	132	13	—	—	691

Max/min °C			Relative humidity %		
Dec	Apr	Absolute	Hour	Mar	Aug
33	41	Jan Apr	0630	23	87
16	26	7 47	1230	11	64

Goias, Central Brazilian Plateau. 15° 58′ S. 536 m													
	J	F	M	A	M	J	J	A	S	O	N	D	Total
°C	23	24	24	25	24	22	22	24	26	26	24	23	—
mm	318	252	259	117	10	8	—	8	59	135	239	242	1646

Max/min °C			Relative humidity %		
Jun	Sep	Absolute	Hour	Aug	Dec
32	34	Jun Sep	0630	64	90
13	18	5 40	1330	40	73

Table 5.3 The regimes are similar, but during the cooler winter months the air at Goias remains relatively humid compared with that at Sokoto, which experiences the dry, dusty Harmattan

soar and the air pressure falls over the outer tropics. They now form part of the inter-tropical convergence zone (ITCZ). Moist air flows in, clouds build up in hitherto clear skies, and periods of heavy thundery rain occur. In the outer tropics, too, mean temperatures tend to fall slightly during the rainy season.

As the ITCZ retreats once more to lower latitudes, the rains gradually cease and the flow of drier air is re-established.

The rainfall amounts, the length of the dry season, and the forms of vegetation vary considerably between the moist low latitude *Af/Aw* transition zone and the borders of the *Aw/BS* climates, where summer temperatures are high, evaporation rapid and the rainfall much less reliable. However, the transition from woodland to plants with xerophytic characteristics, to combat long dry periods is far from regular; variations in soils and micro-climates abound within the vast areas broadly described as 'savanna' (p.116).

In the hottest parts of the year air temperatures rise to well over 40°C, and in the cooler season may reach 25°–20°C during the afternoon; though the night temperature may fall below 15°C. Convection is active through the year. It promotes heavy rain from moist air in the hot season, and swirling 'dust devils' in the more stable air of the drier months.

Fig. 5.8 At Manyara on the East African plateau, beneath the rift valley edge, a game reserve with tall trees and cleared grassy areas stretches away towards a lake. The climatic regime is similar to that of the savanna shown in Fig. 5.9, though the lake provides additional moisture.

Fig. 5.9 A hundred kilometres away on the East African plateau, Maasai herds (**right**) graze a short-grass savanna, and act to prevent tree regeneration. The porous, eroded tephra slopes beyond support few trees; but low trees, bushes, and taller grasses grow on the old solidified lava flows which extend out from the old volcano.

HOT DESERT CLIMATE AND SEMI-ARID CLIMATES

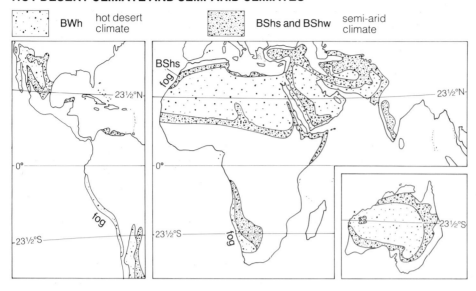

Fig. 5.10 Notice the symbol *s* (summer drought) for semi-arid regions bordering the Mediterranean lands, and *w* (winter drought) for those adjoining the Aw climatic zones.

Fig. 5.7 shows the distribution of *Aw* climates. But most of these areas include extensive uplands with average temperatures below those associated with a typical *Aw* climate.

In India, south-east Asia, and northern Australia the alternating rains and droughts are part of a monsoon system. Yet they involve a poleward migration of the ITCZ during the hot season and have fundamentally similar regimes. But compared with other *Aw* regions, the rain sets in with the sudden arrival of an inflow of moist air.

The term 'savanna': a warning

Savanna describes the tropical grasses which are dominant in the vegetation in the *Aw* regions. This varies from thorn scrub and coarse tussocky grasses in dry areas, through park-land savannas, with scattered trees and bushes, to close woodland and tall elephant grass in moister locations.

But while the vegetation reflects changes in soil moisture, the plant population is also affected by factors which are not primarily climatic. The regeneration of shrubs and trees can be hampered by continual clearance, animal grazing, and frequent fires. It is unwise to attribute a particular form of savanna solely to climate, and most inaccurate to refer to 'a savanna climate'.

5.11 Semi-arid, outer tropical (*BShw*)

These regions are directly affected by air subsidence in the sub-tropics and have a long dry winter season (*w* = winter drought). There is a short rainy period as the ITCZ moves poleward during summer. Annual rainfall varies considerably with fluctuations in the latitudinal movement of the ITCZ. Periods with average rainfall can be followed by many years of exceptional drought. Rain during the hottest months, with evaporation at a maximum, is less effective for plant growth than at other times.

The Sahel countries which stretch across Africa south of the Sahara share an unreliable climatic regime. The danger is that periods of adequate rainfall encourage semi-nomadic people to increase the number of cattle grazing these dry scrub-grasslands. Yet, as in recent decades, prolonged drought increases desertification not only along the edge of the desert, but also in more settled lands to the south, to which many of the semi-nomads migrate. This is discussed further on p.152.

Precarious conditions of this kind are also found in north-central and western Australia, southwest Africa, parts of the Mexican plateau, and in the dry enclave of eastern Brazil.

Table 5.4 *BShw* characteristics

Hall's Creek, North-west Australia. 18° 13′ S. 366 m														Max/min °C				Relative humidity %		
	J	F	M	A	M	J	J	A	S	O	N	D	Total	Jul	Nov	Absolute		Hour	Sep	Jan
°C	30	29	28	26	21	19	18	21	24	28	31	31	—	27	38	Jul	Jan	0930	29	51
mm	137	107	71	13	5	5	5	2	2	13	35	79	460	9	23	−1	44	1530	24	36

Table 5.5 *BW* characteristics

In Salah, Southern Algeria. 27° 12′ N. 280 m															Max/min °C			Relative humidity %		
	J	F	M	A	M	J	J	A	S	O	N	D	Total		Jan	Jul	Absolute	Hour	Dec	Jul
°C	13	16	20	25	29	35	37	36	33	27	19	14	—		21	45	Jan	Jul	0700	65	29
mm	2	3	*	*	*	*	—	2	*	*	3	5	15		6	28	−3	50	1300	38	16

* Less than 2 mm

Table 5.6 *BShs* characteristics

Mosul, Northern Iraq. 36° 19′ N. 222 m															Max/min °C			Relative humidity %		
	J	F	M	A	M	J	J	A	S	O	N	D	Total		Jan	Jul	Absolute	Hour	Jan	Aug
°C	7	9	13	19	23	29	32	32	27	21	15	9	—		12	43	Jan	Jul	0600	92	46
mm	71	78	54	48	18	*	*	*	*	5	48	61	384		2	22	−11	51	1500	64	13

* Less than 2.5 mm

5.12 The hot desert (*BWh*)

The subsiding air of the sub-tropics affects the western parts of the landmasses between latitudes 20° and 25°, and strongly influences territories some degrees north and south of this.

During summer the air over the desert may contain a considerable amount of water vapour, but, with the intense heat, its relative humidity is low. Stable, subsiding air supresses the columns of air rising by convection from the hot surface. Only occasionally do strong updraughts rise to sufficient height for condensation to occur and cumulonimbus to develop. Sometimes such clouds yield rain which evaporates in the relatively dry air before it reaches the ground. But every now and then local torrential downpours cause flash floods to course down dry channels, before soaking into broad depressions. Such channelled water and sheet-wash erode bare surfaces.

Summer shade temperatures may reach 50°–55°C. But by night, radiation losses through the clear atmosphere cause the air temperature to drop sharply, usually to some 20°–24°C. During winter, with a relatively low angle of the noonday sun, air temperatures rise to about 15°–20°C, and at night fall to 5°–10°C, with near-surface temperatures at or below freezing point. Such cooling is apt to produce heavy dew, an important agent of rock weathering.

We have already seen that the western margins are flanked by cool currents and upwelling water, emphasising the aridity, and causing fogs and low stratus cloud.

5.12 Semi-arid: poleward of the hot deserts (*BShs*)

In these areas rain belts associated with mid-latitude depressions affect the borders of the semi-arid lands during winter, bringing occasional heavy storms. The outer limits of the desert are difficult to define on a rainfall basis. Average figures mean little where years of drought may be followed by sudden storms bringing hundreds of millimetres of rain during an hour or so.

The summer months are dry and very hot (*s* = summer drought). But during winter the rainfall is more effective for plant growth. Coarse grasses and xerophytic plants thrive in sufficient quantity to support a limited number of grazing animals on the move.

Such conditions occur over large tracts of north Africa, northern Arabia, eastward to north-west India, and in southern Australia. But a summer rainfall maximum is more common in northern Mexico and the adjoining semi-arid parts of the USA.

Fig. 5.11 The semi-arid landscapes of northern Australia support a surprising amount of wildlife. Here, near Hall's Creek, during the moister summer months tall grasses and small, leafy shrubs surround a broad boab, which resembles the African baobab in having water-storing tissues. Yet, on an average, only 80 mm of rain falls during a nine-month dry period. A comparison with the vegetation in Figs. 5.8 and 5.9 confirms that 'savanna' should never be used climatically.

Fig. 5.12 Here, in the Sinai, what little rain falls comes in the winter months. Even so, plant life is supported mainly by ground-water, replenished by drainage into the valleys. A few deep-rooted trees survive amid the debris left by an occasional flash-flood.

Climates of the middle latitudes

5.13 Warm temperate summer dry: Mediterranean type (*Cs*)

This is experienced on the western side of continents about an axis of latitude 35°. Under the influence of the subsiding dry air of the sub-tropics, the summer is hot, with abundant sunshine and little rain. In winter the cT air is less dominant. There are incursions of air of mP and cP origin, and a number of polar front depressions move in from the west.

In the Mediterranean region itself depressions occasionally form locally. Air movements about such cyclonic systems can draw hot, dusty winds like the Khamsin and Sirocco from the southern deserts, or cold air from mountain sources to bring freezing winds like the Mistral and Bora to the countries of the northern Mediterranean.

Winters, however, are generally mild, though night air temperatures may fall to freezing point or below, and variable weather may bring unusually cold spells. Crops sensitive to cold avoid places which collect or receive an influx of cold air. Frost precautions are usual in fruit-growing districts.

Table 5.7 shows that temperatures near the borders of arid regions differ considerably from those in cool, maritime locations. Annual rainfall also varies a great deal from wetter western coastlands to sheltered inland locations. As winter rain comes mostly from quick-moving fronts and depressions, there are usually long sunny periods. These favour plant growth, especially as the rate of evaporation is less than during summer.

There were once extensive mixed evergreen woodlands about the Mediterranean Sea, climatically suited to the mild growing season and subsequent summer drought. Today, after long human occupation, the characteristic vegetation is the

Table 5.7 Notice the contrasts between coastal stations and those inland, with their higher temperature ranges and lower rainfall

Nicosia, Cyprus. 35° 09′ N. 218 m

	J	F	M	A	M	J	J	A	S	O	N	D	Total
°C	10	10	12	17	22	26	28	28	26	21	16	12	—
mm	74	51	33	20	27	10	*	*	5	23	43	76	364

Max/min °C			Relative humidity %		
Jan	Jul	Absolute	Hour	Jan	Jul
14	36	Feb Jul	0800	85	49
6	21	−5 47	1400	65	29

* Less than 2.5 mm

Lisbon, Portugal. 38° 43′ N. 95 m

	J	F	M	A	M	J	J	A	S	O	N	D	Total
°C	11	11	13	14	17	20	22	22	21	17	14	11	—
mm	84	81	79	61	43	18	5	5	35	79	107	91	686

Max/min °C			Relative humidity %		
Jan	Aug	Absolute	Hour	Jan	Jul
13	27	Feb Jul	0900	83	61
8	18	−2 39	1500	72	46

Stations in different *Cs* regions

Location			*Max/min °C		Total
			Summer	Winter	mm
San Francisco	38° N	16 m	18/12	13/7	510
Fresno	*37° N*	*101 m*	*37/18*	*12/3*	*231*
Perth	32° S	60 m	29/17	17/9	882

Location			*Max/min °C		Total
			Summer	Winter	mm
Valparaiso	33° S	41 m	22/13	16/18	506
Santiago	*33° S*	*510 m*	*29/12*	*15/3*	*358*
Cape Town	34° S	17 m	26/16	17/7	508

italics – inland station   * figures for Jan/Jul

'MEDITERRANEAN CLIMATE'
WARM TEMPERATE WEST COAST - Cs

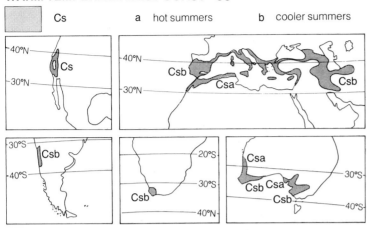

Cs   a hot summers   b cooler summers

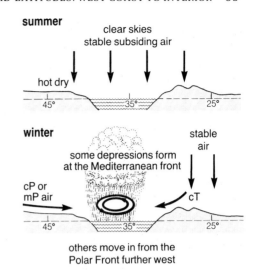

summer

clear skies
stable subsiding air

hot dry

winter

some depressions form
at the Mediterranean front

cP or
mP air

stable
air

cT

others move in from the
Polar Front further west

Fig. 5.13 The contrasts between the *Csb* and *Csa* climates are borne out by the statistics for Lisbon and Nicosia (Table 5.7). The section emphasises that high pressure established over the western sub-tropics extends over these regions during summer. In winter low pressure systems from the Atlantic pass through the Mediterranean, and on eastwards; though there are occasional frontal clashes, with local depressions.

chaparral or **maquis**, described on p.125. In Europe the distribution of the olive tree closely corresponds to the extent of this type of climate.

Table 5.7 emphasises the differences between the marine sub-type of climate in southern Portugal (*Csb*), with its cooler summers, and that of the eastern Mediterranean (*Csa*), with its lower rainfall and greater temperature extremes. Apart from high mountains, countries bordering the Mediterranean include other climatic regions, such as the Spanish steppe-like Meseta, with its colder winters and occasional thundery summer

rain. The plains of northern Italy have heavy rain during summer, and the annual precipitation is high for a *Cs* climatic type.

In Chile and California the western coastlands are affected by cool offshore waters. Each country has a central valley parallel with the coast; this is much sunnier and drier than the sometimes foggy coastal areas. The cooling effect is also evident in south-west Africa and south-western Australia, but less marked.

5.14 Middle latitude interior desert (*BWk*)

The heart of the northern continents, especially central Asia, receives little moist air, particularly where high ranges block any maritime influences. Precipitation is slight and irregular. Most comes from summer convection storms or low-level monsoonal inflows of moist air. Summer convection can also create strong, searingly hot, dry

MID-LATITUDE DESERT

 BW

SEMI-ARID

 BSk

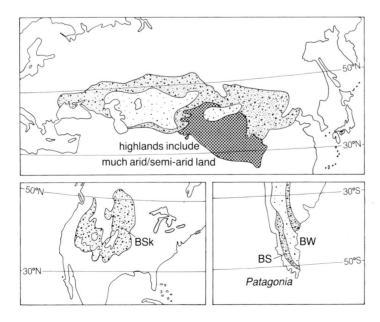

Fig. 5.14 In each of these areas the climate is much modified by relief. In the case of Patagonia, in the latitude of the Roaring Forties, shelter and föhn winds from the Andes are dominant effects.

Table 5.8 The continental interiors

Akmolinsk, Kasak, SSR. 51° 12′ N. 405 m														Max/min °C				Relative humidity %		
	J	F	M	A	M	J	J	A	S	O	N	D	Total	Jan	Jul	Absolute		Hour	Jan	Jul
°C	−19	−18	−12	0	6	17	19	17	11	1	−8	−15	—	−16	26	Jan	Jul	0700	86	67
mm	18	13	13	15	28	43	40	38	25	28	18	15	297	−22	13	−36	35	1300	83	44

Ashkhabad, Turkmen, SSR. 37° 57′ N. 226 m														Max/min °C				Relative humidity %		
	J	F	M	A	M	J	J	A	S	O	N	D	Total	Jan	Jul	Absolute		Hour	Jan	Jul
°C	−1	7	8	15	19	26	29	24	22	15	8	4	—	3	36	Jan	Jul	0700	87	54
mm	25	20	48	35	30	8	3	3	3	13	20	18	226	−4	22	−26	45	1300	69	28

Medicine Hat, Alberta, Canada. 50° 01′ N. 653 m														Max/min °C				Relative humidity %		
	J	F	M	A	M	J	J	A	S	O	N	D	Total	Jan	Jul	Absolute		Hour	Jan	Jul
°C	−11	−10	−2	7	13	17	21	19	13	7	−2	−7	—	−6	29	Jan	Jul	0500	87	78
mm	15	15	15	20	40	61	45	35	28	15	18	18	325	−19	13	−43	42	1100	77	45

Winnemucca, Nevada, USA. 40° 58′ N. 1324 m														Max/min °C				Relative humidity %		
	J	F	M	A	M	J	J	A	S	O	N	D	Total	Jan	Jul	Absolute		Hour	Jan	Jul
°C	−2	1	7	8	13	17	22	21	15	9	3	−1	—	4	32	Jan	Jul	0500	83	47
mm	28	23	25	20	20	15	5	5	10	15	20	25	213	−8	12	−38	42	1200	62	19

Table 5.9 Patagonia as a special case

Sarmiento, Patagonia. 45° 36′ S. 268 m														Max/min °C				Relative humidity %		
	J	F	M	A	M	J	J	A	S	O	N	D	Total	Jul	Jan	Absolute		Hour	Jul	Jan
°C	18	18	14	11	7	3	3	6	8	12	14	16	—	7	26	Jun	Jan	0730	37	52
mm	5	8	8	10	20	20	15	13	13	10	5	8	140	−2	11	−11	37	1330	59	29

dust-laden storms. Air tends to descend into inland basins, which become extremely arid.

These differ from the *BWh* hot desert climates, for the winter is often bitterly cold. In winter the humidity of the cold air may be as low as 20 per cent, when thermal high pressure dominates. Some snow may fall, but generally very little.

Generalisations are of limited value, however, for within Asia and North America arid and semiarid conditions occur through some 20° of latitude. Also the topography varies from high plateaus with outstanding ranges, to low inland basins. In the western parts of Euro-Asia the winter high is less well established, so that maritime influences bring more precipitation.

5.15 Middle latitude semi-arid (*BSk*)

There are enormous tracts of semi-arid country in central Asia and North America, where winters are extremely cold and snow builds up as the months pass. Annual precipitation is of the order of 250–300 mm, but, again, varies with the location, and is most unreliable. Its effectiveness for plant growth depends on the rate of evaporation. Evaporation considerably exceeds precipitation, especially in the lower latitudes. Summer rain comes with influxes of moist air, usually in brief thundery downpours; but the hot surfaces dry rapidly.

Grass is dominant in the dry steppe vegetation, often with bare soil between plants. This has encouraged extensive grazing and un-irrigated agriculture; but under these climatic conditions this has also led to serious soil erosion.

In Patagonia semi-arid conditions extend from the immediate rain-shadow of the Andes to the east coast. Air from the west, warmed on descent, is relatively dry, and the cold offshore Falkland Current helps to maintain the aridity; though, even in the south, mean winter temperatures are above freezing. Frequent strong winds dry the low tussocky grassland.

5.16 Humid sub-tropical (*Ca*)

Climates with no distinct dry season (*Caf*) are found in the eastern parts of the continents in the

Table 5.10 *Caf* climates – no dry season

Nanjing, Eastern China. 32° 03′ N. 16 m															Max/min °C			Relative humidity %		
	J	F	M	A	M	J	J	A	S	O	N	D	Total		Jan	Jul	Absolute	Hour	Jan	Jul
°C	2	4	9	14	20	24	28	27	22	18	11	4	–		6	31	Jan	Jul	0600	43	88
mm	40	51	76	102	81	183	206	117	94	51	40	30	1072		−2	24	−13	40	1400	29	75

Port Macquarie, Eastern New South Wales. 31° 38′ S. 19 m															Max/min °C			Relative humidity %		
	J	F	M	A	M	J	J	A	S	O	N	D	Total		Jul	Jan	Absolute	Hour	Aug	Jan
°C	22	22	21	18	16	13	12	13	15	17	19	21	–		18	26	Jun	Feb	0900	75	75
mm	140	178	162	165	142	120	110	84	96	89	94	127	1507		7	18	−1	41	1500	65	75

Charleston, South Carolina, USA. 32° 47′ N. 3 m															Max/min °C			Relative humidity %		
	J	F	M	A	M	J	J	A	S	O	N	D	Total		Jan	Jul	Absolute	Hour	Jan	Jul
°C	10	11	14	18	23	26	27	27	25	20	14	11	–		14	31	Feb	Jul	0730	81	79
mm	74	84	86	71	81	120	186	167	130	81	59	71	1205		6	24	−14	40	1200	64	67

Stations in different *Caf* regions

Location		*Max/min °C		Total		Location		*Max/min °C		Total
		Summer	Winter	mm				Summer	Winter	mm
Guangzhou (China)	23° N	33/26	17/9	1644		Durban (S. Africa)	31° S	27/21	23/11	1009
Rockhampton (Aus)	23° S	32/22	23/10	993		Osaka (Japan)	35° N	31/23	8/0	1336
Santos (Brazil)	24° S	29/22	23/16	2238		Buenos Aires (Arg)	35° S	29/17	14/5	950
Mobile (Alabama)	31° N	32/33	16/7	1577		Springfield (Missouri)	37° N	31/20	6/−4	1069

* Figures for January or July according to the hemisphere. Notice the contrast in winter temperatures between the northern and southern *Caf* regions.

Fig. 5.15 *Ca* climates – with variations.

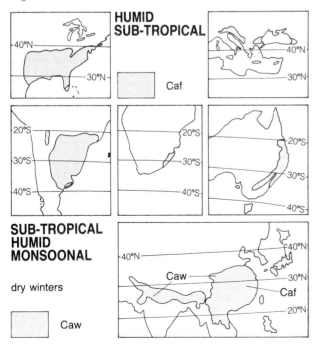

HUMID SUB-TROPICAL

Caf

SUB-TROPICAL HUMID MONSOONAL

dry winters

Caw

Caw

Caf

Caf

lower middle latitudes and sub-tropics. mT air moves from the western parts of the oceans over the land areas, especially during summer when the interior is hot. Though rain falls throughout the year, there is usually a summer maximum; for tropical downpours, with thunderstorms, develop in the oppressively humid air, with its mean temperature of some 22°–30°C. Conditions are drier towards the interior of the landmasses. Hurricanes affect the eastern seaboard, and are most frequent in autumn.

In all these regions there is greater stability in winter, when incursions of maritime air pass over warm, rather than hot, land surfaces. In south-east Australia with its well-distributed rainfall through the year, and in Argentina, the winters are warm and mild, with occasional frontal rain belts. But in eastern China air from the cold Asiatic interior brings much lower winter temperatures. Shanghai's January mean of 3°C compares with July averages of 10°C in Buenos Aires and 11°C in Sydney.

Winters are generally mild in south-eastern USA, though surges of cold polar air conflict with

Table 5.11 *Caw* climates – dry winters

New Delhi, Northern India. 28° 35′ N. 218 m														Max/min °C		Absolute		Relative humidity %		
	J	F	M	A	M	J	J	A	S	O	N	D	Total	Jan	May	Absolute		Hour	Nov	Aug
°C	14	17	22	28	33	33	31	30	29	26	20	15	—	21	41	Jan	May	0800	51	80
mm	23	18	13	8	13	74	181	172	117	10	2	10	640	7	27	−1	46	1630	31	64

Chongqing, Central China. 29° 33′ N. 230 m														Max/min °C		Absolute		Relative humidity %		
	J	F	M	A	M	J	J	A	S	O	N	D	Total	Jan	Aug	Absolute		Hour	Nov	Aug
°C	7	10	15	19	23	26	29	30	25	19	14	10	—	9	35	Jan	Aug	Daily	91	76
mm	15	20	38	99	142	181	142	122	142	112	48	20	1032	5	25	−2	44	mean		

mT air and periods of steady frontal rain occur. Colombia, South Carolina, averages 8°C in January, and parts of the south-east have a slight winter maximum rainfall.

Small parts of southern Europe have climatic characteristics which place them in the *Caf* category. For instance, the plains of northern Italy and the lower Danube basin receive precipitation through the year, but with a summer maximum.

There is considerable regional variety in the temperature and rainfall figures, even for places within the humid sub-tropics, as Table 5.10 shows for the *Caf* climatic regions.

The south-central inland parts of eastern China, like the plains of northern India, have a *Caw* climate, with very dry winters (Table 5.11).

5.17   Cool temperate humid (maritime) *(Cb, Cc)*

This is typical of the western parts of the continents in the middle latitudes, where eastward-moving depressions and prevailing westerlies bring oceanic influences over the land areas. Marine controls modify latitudinal ones, and in winter tend to overshadow them, as the isotherms in Fig. 4.4 clearly show. They affect lands from latitude 40° to 60° and over.

The extent of penetration of maritime influences

Fig. 5.16 These temperate *Cbf* regions, with rain in all seasons, include those with cool winters (I), and others in South Africa, eastern mainland Australia, and the North Island of New Zealand with considerably warmer winters (II).

into land areas depends largely on relief. In north-central Europe oceanic influences are carried far inland across extensive lowlands; whereas in Canada and Chile high mountains restrict this type of climate to narrow western coastlands. The converse is also true; continental influences are more apparent in western European countries than in other west coast regions of this type.

The eastward procession of lows, with associated fronts, ridges and troughs make for changeable weather. Maritime influences keep the mean summer temperatures several degrees below the latitudinal mean, mostly of the order of 13°–18°C, depending on the latitude and distance from the ocean. Long drought periods are rare, though extensions of sub-tropical high pressure occasionally bring spells of unusual heat, and even droughts.

Maritime influences are strong in winter. In Europe air from the relatively warm Atlantic keeps western coasts much milder than the interior. The ports of northern Norway are some 15 C° warmer than the average for the latitude and remain open for shipping; even those within the Arctic. However, bitterly cold cP air occasionally covers western Europe in mid-winter; though this can change in a matter of hours to warm, damp conditions, as westerlies re-establish themselves, or as a warm front arrives. The continental influences increase away from the western coasts. This is noticeable even in so narrow and westerly a country as Britain.

There is precipitation throughout the year,

COOL TEMPERATE HUMID CLIMATES - Cbf and Cc

West Coast Maritime - rain in all seasons

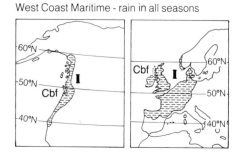

I cool winters   II warm winters

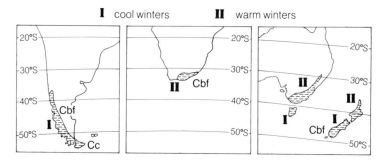

though amounts vary from place to place, and depend very much on relief. In Europe windward western hills can receive 5000 mm a year, while rain-shadow areas a short distance inland have as little as 500 mm. It is chiefly associated with fronts which cross land areas; but high relief causes additional uplift and increases precipitation. Western parts also tend to a winter maximum, often with high rainfall in autumn. During winter, depressions are more frequent; though high pressure to the east may slow their passage across western parts, so that cloudy skies and periods of steady, light rain may persist for long periods. Normally snowfall does not last long over the lowlands; but its duration increases towards the continental interiors.

Depressions are less numerous in summer, but the systems move more easily towards the continental low pressure areas, where rainfall is markedly greater at this time of year; not because there are more rainy days, but because convection updraughts over heated lowlands bring thundery storms.

Anticyclones produce hot, sunny weather in summer; but in winter bring long periods of frost, or persistent fog.

In some *Cb* regions there is little likelihood of well-established anticyclones, especially in the southern hemisphere, where the narrow land masses hardly interrupt the air flows across the expanses of ocean. In New Zealand depressions move rapidly eastward; so short rainy periods are followed by clear, bright weather with much sunshine. In summer, however, the North Island receives occasional cyclonic storms of tropical origin. Southern Chile experiences strong westerlies and a succession of eastward-moving depressions throughout the year.

In these regions the natural lowland vegetation is forest, though parent soil material has considerable influence on the actual plant associations (p.100); and in western Europe, especially, little *natural* forest remains. Conifers tend to replace deciduous or mixed forest on highlands and in the poleward parts of these regions.

Table 5.12 *Cbf* (I) regions in the northern hemisphere

Shannon, Western Ireland. 52° 41′ N. 2 m

	J	F	M	A	M	J	J	A	S	O	N	D	Total
°C	5	6	7	9	11	14	15	16	14	11	8	6	—
mm	96	76	51	56	61	54	79	76	76	86	107	110	927

Max/min °C				Relative humidity %		
Jan	Aug	Absolute		Hour	Jan	Aug
8	20	Jan	Jul	0630	89	91
2	12	−11	31	1730	88	77

Berlin, Germany. 52° 27′ N. 57 m

	J	F	M	A	M	J	J	A	S	O	N	D	Total
°C	−1	0	4	8	13	16	18	17	14	9	4	1	—
mm	48	33	38	43	48	59	79	56	48	43	43	48	609

Max/min °C				Relative humidity %		
Jan	Jul	Absolute		Hour	Jan	Jul
2	23	Feb	Jul	0700	89	79
−3	13	−26	36	1400	81	55

Prince Rupert, British Columbia. 54° 17′ N. 52 m

	J	F	M	A	M	J	J	A	S	O	N	D	Total
°C	1	2	4	6	9	12	13	14	12	9	5	2	—
mm	249	193	213	170	135	105	122	130	196	310	313	287	2421

Max/min °C				Relative humidity %		
Jan	Aug	Absolute		Hour	Jan	Aug
4	18	Jan	Jun	0400	83	96
−1	11	−19	31	1600	78	77

Table 5.13 *Cbf* regions in the southern hemisphere

Cabo Raper, Southern Chile. 46° 50′ S. 40 m

	J	F	M	A	M	J	J	A	S	O	N	D	Total
°C	11	11	10	9	8	7	6	6	7	8	9	10	—
mm	198	147	181	196	191	201	242	191	142	178	170	178	2212

Max/min °C				Relative humidity %		
Jul	Jan	Absolute		Hour	Jul	Jan
8	14	Jun	Mar	0700	83	85
3	8	−2	22	1400	82	82

Invercargill, South Island, New Zealand. 46° 26′ S. 4 m

	J	F	M	A	M	J	J	A	S	O	N	D	Total
°C	14	14	13	11	8	6	5	7	9	11	12	13	—
mm	107	84	102	105	112	91	81	81	81	105	107	102	1155

Max/min °C				Relative humidity %		
Jul	Jan	Absolute		Hour	Jul	Jan
9	19	Jul	Jan	Mean daily	83	76
1	9	−7	32			

Fig. 5.17 West of New Zealand's Southern Alps the relative warmth and humidity favour trees festooned with epiphytes and vines, tree ferns, and a thick shrubby undergrowth.

Fig. 5.18 East of the Southern Alps introduced trees, sown pastures, and grassy hillsides, with low shrubs above, contrast with the western forests (Fig 5.16), and show the limitations of broad classifications.

5.18 Continental humid–cold winter (*Da, Db*)

There is no broad land area south of latitude 40°S, other than Antarctica, so these climates occur in northern landmasses. They extend from eastern Europe, from the Atlantic coast of North America, and from Asia's Pacific coast far into the continental interiors, mainly between 40°–60°N.

Because of this great spread, the symbols *Da* are used to indicate warm summers and *Db* the shorter cooler summers (but with the advantage of longer daily periods of insolation).

In Euro-Asia these lands are affected in winter by the persistent flow of cold air from the thermal high pressure sources of central Asia. Precipita-

Fig. 5.19 These regions extend between latitudes 40°N and 60°N, so that the southern parts have much warmer summers (*Da*). Many parts of the interior have less than half the rainfall of the areas with maritime influences.

tion occurs where the cold cP air meets maritime air; from the somewhat moister eastern Europe it decreases towards the dry interior of central Asia. In eastern Asia winter precipitation from the cold, dry, often dusty air is small.

In North America the more northerly parts of the interior experience long cold spells during winter, and waves of cold cP air move southward and eastward from time to time, displacing mT air and causing a sharp fall in temperature. Winter weather is changeable in the rest of the American *Da* and *Db* regions. Depressions are apt to form in the central and eastern parts; as they move eastward, they occasionally bring periods of heavy snow. In the northern interior the precipitation is lighter, but snow lies for long periods and lowers temperatures by reflecting insolation; though the snow cover helps to prevent deep freezing of the ground.

HUMID CONTINENTAL WITH COLD WINTERS - Da and Db

 Da warm summers

 precipitation 250-500mm p.a.

Db cool summers

remainder of shaded area over 500mm p.a.

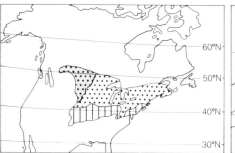

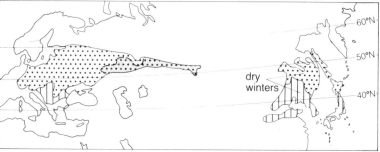

Table 5.14 *Da* climates. East Coast Climates – very warm to hot summers

Boston, Massachusetts, USA. 42° 22′ N. 38 m														Max/min °C		Absolute	Relative humidity %		
	J	F	M	A	M	J	J	A	S	O	N	D	Total	Jan	Jul		Hour	Jan	Jul
°C	−2	−2	2	8	14	19	22	21	17	12	6	−1	—	2	27	Feb Jul	0800	72	71
mm	71	84	96	89	79	81	84	71	81	84	71	86	1036	−7	17	−28 40	1200	63	66

Inchon, South Korea. 37° 29′ N. 70 m														Max/min °C		Absolute	Relative humidity %		
	J	F	M	A	M	J	J	A	S	O	N	D	Total	Jan	Aug		Hour	Jan	Jul
°C	−3	−2	6	11	16	19	24	25	20	14	6	2	—	1	29	Jan Aug	0530	73	95
mm	20	18	30	66	84	99	277	223	110	40	40	28	1036	−7	22	−21 37	1330	57	76

In all these regions summer temperatures are high for the latitudes. In both America and Euro–Asia there is generally greater precipitation during summer, when convection is at a maximum, and moist unstable air moves more freely into the landmass; though air from the eastern oceans moves in against the general air flow in these latitudes. Thundery storms develop in the mT air, giving short torrential downpours.

Depressions also bring rain in summer, especially in North America, but they are less frequent in winter. Autumn and spring seasons are short, with tendencies towards rain in spring, but dry, clear weather in autumn.

The overall climatic differences between the various parts of the *Da*, *Db* regions are great. In general, the precipitation decreases poleward and towards the interior. The eastern parts of North America do not experience the summer and win-ter differences found in eastern Asia, though the drier interiors have a pronounced summer maximum.

The more southerly and more maritime locations have milder winters, and the summers are particularly hot in the southern parts of the interior.

Though distinctions between the *Da* and *Db* climates are necessary, and valid overall, no clear boundary exists on the ground. Also the vegetation associated with these regions varies from place to place. In the heart of the prairie grassland, deciduous trees follow the streams and river valleys. In general, tree cover increases towards the more humid parts, and southern deciduous hardwoods give way poleward to mixed forest, and then to coniferous softwoods. But these are broad generalisations; for variations in relief and soils modify the local vegetation.

Table 5.15 *Db* climates. Landmass interior

Minsk, White Russian SSR. 53° 54′ N. 27° 33′ N. 230 m														Max/min °C		Absolute	Relative humidity %		
	J	F	M	A	M	J	J	A	S	O	N	D	Total	Jan	Jul		Hour	Jan	Jun
°C	−8	−7	−2	4	12	15	17	16	11	5	−1	−6	—	−6	21	Jan Jul	0700	91	76
mm	35	38	33	38	51	71	76	79	40	38	38	43	582	−11	12	−33 33	1300	86	58

Kazan, RSFSR. 55° 47′ N. 49° 08′ E. 80 m														Max/min °C		Absolute	Relative humidity %		
	J	F	M	A	M	J	J	A	S	O	N	D	Total	Jan	Jul		Hour	Jan	Jul
°C	−15	−13	−7	3	12	16	19	11	7	1	−6	−10	—	−13	24	Jan Jul	0700	88	75
mm	20	15	18	20	33	61	59	51	43	40	30	23	414	−18	14	−43 38	1300	85	49

Regina, Saskatchewan, Canada. 50° 26′ N. 570 m														Max/min °C		Absolute	Relative humidity %		
	J	F	M	A	M	J	J	A	S	O	N	D	Total	Jan	Jul		Hour	Jan	Jul
°C	−18	−17	−9	3	11	16	18	17	11	4	−6	−14	—	−12	26	Jan Jul	0530	91	88
mm	13	8	18	18	45	84	61	45	33	23	15	10	374	−24	11	−49 42	1200	86	54

5.19 Sub-arctic (*Dc, Dd*)

These lie poleward of the middle latitude continental climatic regions. Their more northerly parts, especially, are characterised by long, extremely cold winters, short summers, and large annual temperature ranges. For the *Dd* climatic region the mean temperature of the coldest month is below −38°C. The poleward limit is taken as the 10°C isotherm for the warmest month. Below this mean temperature tree growth is inhibited, and tundra is the chief form of vegetation. Extensive softwood coniferous forest (taiga) is typical of the *Dc* climatic regions.

During the long winters the few hours of low intensity insolation cannot make up the radiant energy losses, so that in the centre of the continents, where the absolute humidity is low, temperatures fall to below −50°C. Cold, dense air masses build up intense thermal highs, from which air surges at times into lower latitudes and towards the coasts.

Central Siberia is excessively cold in winter; in fact the *Dd* sub-division applies only to parts of north-eastern Siberia, where Verkhoyansk has recorded −68°C. In southern Siberia the mean figure is of the order of −20° to −25°C; close to the mean January figure of −27°C in Churchill, Canada.

During the short summer the temperature climbs rapidly, for although the altitude of the noonday sun is not great, there are many daylight hours. Mean monthly temperatures rise to over

Fig. 5.20 The mean temperature of the coldest month is below 0°C. For the *E* Group climates that of the warmest month is under 10°C.

NORTHERN POLAR LANDS - Dc and Dd: ET and EF

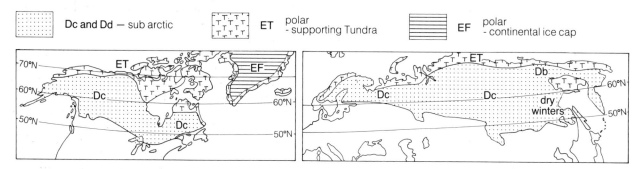

Table 5.16 *Dc* climates. Note that Verkhoyansk is classified as a *Dd* climate, for despite the extremes of cold during winter, air temperatures during summer are surprisingly high

Gällivare, Sweden. 67° 08′ N. 365 m														Max/min °C				Relative humidity %		
	J	F	M	A	M	J	J	A	S	O	N	D	Total	Jan	Jul	Absolute		Hour	Jan	Jul
°C	−11	−12	−8	−2	6	11	15	12	6	−1	−1	−10	—	−7	21	Jan	Jul	0830	84	68
mm	43	28	25	33	38	59	76	74	54	59	45	38	569	−16	9	−42	34	1430	83	52

Bogolovsk, RSFSR. 59° 45′ N. 59° 01′ E. 192 m														Max/min °C				Relative humidity %		
	J	F	M	A	M	J	J	A	S	O	N	D	Total	Jan	Jul	Absolute		Hour	Jan	Jul
°C	−21	−16	−9	−3	7	13	16	14	8	−1	−12	−19	—	−17	21	Jan	Jul	0700	85	77
mm	15	18	18	25	45	69	84	84	43	23	23	20	467	−24	12	−42	34	1300	79	59

Churchill, Manitoba, Canada. 58° 47′ N. 13 m														Max/min °C				Relative humidity %		
	J	F	M	A	M	J	J	A	S	O	N	D	Total	Jan	Jul	Absolute		Hour	Dec	Jul
°C	−28	−26	−21	−10	−1	6	12	11	5	3	−15	−24	—	−24	18	Jan	Jul	0600	97	88
mm	13	15	23	23	23	48	56	69	59	35	25	18	406	−33	6	−50	36	1200	93	71

Dd climate

Verkhoyansk, RSFSR. 67° 34′ N. 133° 51′ E. 100 m														Max/min °C			Relative humidity %		
	J	F	M	A	M	J	J	A	S	O	N	D	Total	Jan	Jul	Absolute	Hour	Jan	Jun
°C	−51	−45	−32	−15	0	12	14	9	2	−13	−38	−47	—	−48	19	Feb Jul	0700	70	62
mm	5	5	3	5	8	23	28	25	13	8	8	5	135	−53	8	−68 37	1300	70	45

16°C, and actual midday readings may approach 30°C. Again the temperatures vary with location; but everywhere has at least one summer month with a mean of 10°C or over.

Most precipitation falls in summer. But even a light winter snowfall may lie for long periods with little melting or evaporation loss, so that snow gradually accumulates through the winter, until the spring melt. The interior of the *Dc* regions receives some 250–500 mm a year; but eastern Canada and north-western Europe have rather more. The *Dd* area has less than 25 mm. Falls are very light in the heart of northern Siberia, though sufficient moisture may be retained to support tree growth.

High latitude climates

5.20   Polar-supporting tundra (*ET*)

This lies between the zone of permanent frost with its cover of ice and snow, the ice-caps where no vegetation grows, and the poleward limits of the sub-Arctic climates. Here no month has a mean temperature above 10°C.

Summers are cool and short, with average temperatures above freezing point for only 2–4 months. But the sun is above the horizon for most of the 24 hours, so that the diurnal range is small. Many days are pleasantly warm, some very warm.

In winter, with no insolation to counter continuous radiation losses for months on end, temperatures fall rapidly and continue to fall, to a bitter cold. Snow remains until May, when melt-water, unable to drain through the permanently frozen sub-soil, produces swampy conditions.

Most precipitation comes during summer; and under the anticyclonic conditions of the continental winter the snow is usually dry and powdery. In continental locations the annual total seldom exceeds 250 mm.

Some tundra regions receive maritime influences, notably those relatively near the coasts of north-western Europe. There the average temperatures are much higher than in the interior; and summer precipitation tends to be higher, as it does in the Labrador peninsula.

The nature and abundance of the vegetation often depends on micro-relief features and local micro-climates. There is usually a mat of plants, including heaths and stunted trees in the warmer parts, but often only mosses and lichens on rocks bordering the ice-caps. The life cycle of many plants lasts but a few months. A carpet of flowers gives way to bright berries before the short autumn.

Table 5.17 *ET* climates – with all monthly mean temperatures below 10°C

Ruskoye Ust'ye, RSFSR. 71° 01′ N. 149° 29′ E. 6 m														Max/min °C			Relative humidity %		
	J	F	M	A	M	J	J	A	S	O	N	D	Total	Jan	Jul	Absolute	Hour	Jan	Jul
°C	−39	−38	−32	−24	−8	4	9	7	0	−14	−27	−35	—	−37	13	Jan Jul	0700	81	79
mm	5	5	5	3	10	20	28	28	18	8	8	8	146	−40	6	−52 32	1300	82	81

Godthaab, South-west Greenland. 64° 11′ N. 51° 43′ E. 20 m														Max/min °C			Relative humidity %		
	J	F	M	A	M	J	J	A	S	O	N	D	Total	Jan	Jul	Absolute		Jan	Jun
°C	−10	−10	−7	−4	1	7	7	6	4	−1	−5	−8	—	−7	11	Jan Jul	Daily	85	92
mm	35	43	40	33	43	35	56	79	84	64	48	38	598	−12	3	−29 24	mean		

Hebron, North-east Canada. 58° 12′ N. 62° 21′ W. 15 m														Max/min °C			Relative humidity %		
	J	F	M	A	M	J	J	A	S	O	N	D	Total	Jan	Aug	Absolute		Jan	Jul
°C	−21	−21	−14	−8	0	4	8	9	5	−1	−7	−16	—	−17	13	Jan Jul	Daily	82	85
mm	23	18	23	28	40	54	69	69	84	40	28	15	491	−24	4	−41 31	mean		

Table 5.18 *EF* – ice-cap climates

Eismitte, Central Greenland. 70° 53′ N. 40° 42′ W. 3000 m

	J	F	M	A	M	J	J	A	S	O	N	D	Total
°C	−36	−47	−40	−31	−21	−17	−12	−18	−22	−36	−43	−38	—
mm	15	5	8	5	3	2	3	10	8	13	13	25	110

Max/min °C				Relative humidity %	
Feb	Jul	Absolute		Feb	Jul
−41	−7	Mar	Jul	Daily mean	77	86
−53	−17	−65	−3			

Little America, Ross Sea, Antarctica. 78° 34′ S. 163° 56′ W. 9 m

	J	F	M	A	M	J	J	A	S	O	N	D
°C	−7	−16	−21	−29	−31	−27	−38	−36	−40	−26	−19	−7

Max/min °C				Relative humidity %	
Sep	Jan	Absolute		Jun	Mar
−34	−4	Sep	Dec	Daily mean	88	76
−46	−9	−59	6			

South Pole, Antarctic Continent. 2956 m

	J	F	M	A	M	J	J	A	S	O	N	D
°C	−41	−49	−57	−57	−56	−58	−62	−59	−55	−45	−35	−39

Absolute minimum at Vostok (78° S): −88.3°C (1960).

5.21 Polar ice-caps and oceans (*EF*)

The characteristics of the Arctic and Antarctic climates are discussed in some detail on p.47, in particular their roles in maintaining the global atmospheric energy balance.

With permanent international bases maintained in the Antarctic, with inter-related scientific projects and exchanges of scientific information, knowledge of the sub-glacial continental structure, of past climatic behaviour revealed by ice, and of conditions in the upper atmosphere, is throwing light on the whole global ecosystem.

Possible causes of a general warming of the earth's atmosphere are discussed on p.151. The effects of atmospheric warming on the ice sheets is as yet uncertain; but the Antarctic ice sheets contain enough water to raise sea levels by 55 m. So an understanding of the relationships between atmosphere and the ice sheets is crucial; for even melting on a lesser scale might flood coastal areas, affect oceanic circulation, and thus trigger other climatic changes.

5.22 Zonal patterns: an overall view

The Meteosat image (Fig. 5.21) not only shows active weather systems on a May day but serves to summarise many of the zonal climatic features described above. In broad terms there are:

1 Clear skies associated with subsidence over the deserts of Africa and the Middle East.
2 A convection cloud belt with storms in the inter-tropical convergence zone in equatorial latitudes, particularly over the western North Atlantic.
3 Moist air creating cloud over many parts of central Africa; it extends northward over East Africa, influenced by the developing summer monsoon in south-east Asia.
4 Moist air is also forming clouds as it moves into southern parts of West Africa; but the interior remains absolutely cloudless.
5 Lines of cloud and a broader rain area are moving westward in the Trade wind zone in the North Atlantic, while the eastern ocean remains under the influence of subsiding air.
6 More stable conditions exist over central and western South Atlantic, and in the lines of cumulus in the south-east Trades open rings of cloud form about a clear centre. This contrasts with the closed cells and high sheets of cloud in the less stable air to the south-east.
7 In temperate latitudes cloudy conditions characterise the zone of the westerlies over the

Meteosat image supplied by the European Space Agency

Fig. 5.21 A Meteosat view of a full earth disc in May, recorded by a geostationary satellite, 35 950 km up, at the intersection of the Greenwich meridian and the equator. A system of Meteosats is located above the equator at 70° longitude intervals. They also sense in the infra-red (p.169), acquiring data for vegetation, soil, and oceanographic studies.

southern oceans, with swirls about the centre of depressions. In the South Atlantic a belt of storms lies where cold air to the west of the low pressure system meets warm, moist air in the sub-tropics.

8 In the northern hemisphere, high pressure with clear air is established over western Europe; thick belts of cloud mark fronts along the flanks. In mid-Atlantic a narrow ridge separates the system affecting western Europe from clouds associated with a depression near Iceland.

6

EXCHANGES OF ENERGY AND MATTER

The many variable climatic elements described in the previous chapters may be regarded as external independent variables of soil-vegetation systems. Such open systems, of whatever scale, from a lichen-covered rocky slope to an extensive rain-forest, exchange energy and matter with their environment and involve cyclic exchanges between plants and soils. So before looking at various types of soil and particular plant communities in detail, it is as well to consider the exchanges of energy and matter involved in the ecosystem as a whole.

6.1 Energy flow and cycling of matter in the biosphere

The sun is the prime external source of energy for the biosphere. Most plants are **autotrophic**; that is they synthesise the organic matter for their various organs from inorganic molecules through photosynthesis. They are **producers**, and are consumed by innumerable species of animal life, which are therefore **consumers**.

Fig. 6.1 shows the way that matter is cycled in this food chain, returning to plants through the actions of micro-organisms which decompose dead and excreted matter, **decomposers**. It also indicates the flows of energy, with losses at each stage. Plants, of course, also obtain nutrients from inorganic sources, and these, too, pass through the food chain, with discards at each **trophic level** (stage in the food chain).

6.2 Plant-soil nutrient exchanges

The biological molecules of all living organisms contain chemical elements besides oxygen, carbon and hydrogen. Some, such as nitrogen, phosphorus, sulphur, calcium, sodium and chlorine, may be present in relatively large quantities; many

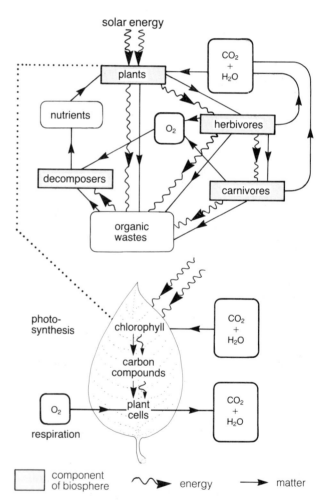

Fig. 6.1 Solar energy activates autotrophs, so that energy and matter together pass through a cycle, with atmospheric controls.

others are contained as minute, but necessary amounts, essential for plant development or animal well-being.

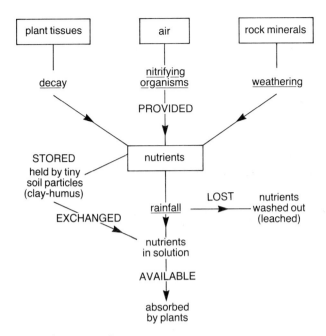

Fig. 6.2   Sources of nutrients.

The plant-soil nutrient system involves chemical actions by which potential nutrients can be absorbed into the plants; nutrients either released to the soil by rock weathering or made available from organic or inorganic sources by the action of micro-organisms in the soil.

Some soil particles have the special property of retaining nutrients until they may be absorbed through plant roots in solution, defying the action of excess water and dilute acids draining through the soil to remove them (p.83). Fig. 6.2 summarises the processes by which nutrients are made available to plants.

6.3   Nutrients, energy and the location of plant communities

Each plant or animal species requires a certain proportion of various elements. A *deficiency* of a particular element may prevent cells functioning properly, or cause death. Thus most plants require a specific quantity of inorganic nitrogen for

healthy growth; yet they also react unfavourably to *excess* – one of the reasons why nitrogen fertiliser or animal manure washed into streams may prove harmful to living things, as the concentration increases.

Plant species are able to flourish in places where the required elements are present in amounts they can tolerate; and animal species adapt to locations where plants and animal prey can provide the necessary nutrients. Thus a food chain, or food web develops, through which energy flows.

Only a small proportion of the energy absorbed by a plant, and used in processes of conversion of elements to organic mass, is passed through **the food chain**, and at each subsequent trophic level only about one tenth of that received is passed on (Fig. 6.3).

Consider the many plants which make up a form of vegetation, say a woodland: each species will compete with others for light, nutrients, water resources, root space, and so on. This will affect the composition of the vegetation, which will also depend on the size, growth-rate, and fertility-rate of the species concerned. Eventually a balanced community may develop, with the various species adapted to their own particular niches; thus achieving **a state of maturity**.

Animals, too, will be involved in this, as the extent of the food chains, and their effects on plants, through such things as seed distribution and fruit consumption, will affect the balance of the ecosystem and the time taken for a biome to become mature.

A **biome** is a complete ecological unit covering a fairly large area and made up of those plants and animals (its **biomass**) which are adapted to the existing physical and climatic conditions. It is affected therefore by the solar input; and as this varies with latitude, there are recognisable biomes, such as those of a rainforest, or savanna, or desert, which are markedly zonal (pp.111–122). As with climatic types, however, the contrasting influences of large landmasses, ocean areas, and varying altitudes, modify this purely zonal distribution.

Fig. 6.3   Energy losses occur at each trophic level.

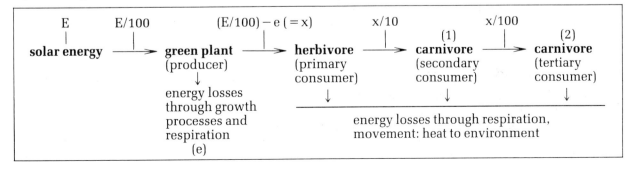

6.4 Interference with soil–vegetation systems

The rate at which the rapidly increasing human population is disturbing mature ecosystems, and the ignorance of so many people about the consequences of their actions, are alarming. So far we have considered the interrelationships of climatic elements in some detail, and can appreciate how fluctuations in one may cascade through so many others: changes in sea surface temperature affect rainfall, which affects energy inputs into upper air streams, which affects global energy transfers, which affects air subsidence – and so affects surface temperatures and air movements in far distant places. In the following chapters we look at the characteristics of soils, plants, and soil–vegetation systems at various scales, and in sufficient detail to draw attention to *their* close interdependence and the possibilities of similar cascades of effects through human interference.

Fig. 6.4 Moorland vegetation shows the importance of recycling the break-down products of organic material. Here the sandstones release few nutrients. The plants grow in acid peaty soil. Heathers are favoured by high humidity, though not by waterlogged conditions, but tolerate acidic soils. The rain and snow bring some nutrients, but wash others from the soil. In this ecosystem most of the nutrients are retained in the plants and slowly rotting litter. When heath is burnt, providing the fire is not too intense, chemical nutrients are released and become a major input to the soil, stimulating new growth.

However, many of these uplands were not always heath or peaty moors. Heathers and other dwarf shrubs have only been able to compete successfully once former woodland was cleared, and controlled grazing by cattle, sheep, and deer help prevent its regeneration. The herbivores derive nutrients from younger plants and supply manure to the system.

7

SOIL FORMATION

7.1 Soil components

Soil is a material consisting of mineral and organic matter in the form of solids, liquids and gases. Here the term is taken to include both the top, earthy layers, in which most of the plant roots are growing, and the sub-soil beneath, which includes bulky material weathered from the parent rock, and may be considerably deeper than overlying earth. Fig. 7.1 shows these simple horizons as a three-dimensional **pedon**. This is a soil unit, with a minimum area of 1–10 m², large enough for the particular properties of each horizon to be described (as in 7.9).

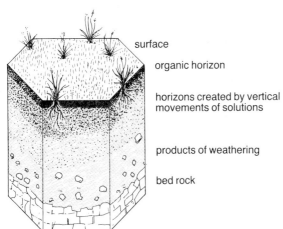

surface

organic horizon

horizons created by vertical movements of solutions

products of weathering

bed rock

Fig. 7.1 A pedon, with visible layering created by physical and chemical changes as solutions soak downward, or are drawn upward to a surface layer rich in decomposed organic matter.

Inorganic particles derived from the break-up of rocks make up the bulk of most soils, though a peaty soil may have a greater proportion of organic matter. The organic content consists of decaying plant and animal matter and the substances resulting from that decomposition. The latter, together with soluble inorganic substances, may dissolve, and form part of the soil liquid content. Living organisms, such as worms, insects and bacteria, affect the composition of the soil in a number of ways.

The size of the material varies from large stones and gravels, through small sand grains, to minute particles of silt and clay, some less than one-thousandth of a millimetre across. Spaces between the particles are filled with gases; mostly those of the atmosphere, but including some produced by bacterial and chemical action.

7.2 Inorganic contents

The parent rock is broken down by mechanical or chemical weathering into fragments of inorganic substances, which come to form a large part of the soil cover. In any place the inorganic matter may have been derived from the local bed-rock or else transported from other sources; and so may be a mixture of materials of various origins.

Even where little transportation has occurred, the parent rock does not alone determine the nature of the soil. Many processes are involved in creating a mature soil, and in certain circumstances soil types may be virtually independent of the parent rock. There are also a number of different soils which have derived inorganic materials from the same bed-rock, but which have been formed under different conditions of climate, relief, or drainage.

Naturally, in a complex substance such as a soil there is a great variety of components, with varying physical and chemical properties. Among the commonest solids resulting from rock disintegration are silica, which in the form of quartz is very resistant to chemical weathering, and various silicates. Together these make up over half the mass of the earth's crust, often as complex molecules combined with aluminium, iron, and potassium oxides. Compounds of calcium and sodium are also numerous.

Some soils contain only small amounts of certain elements; yet these **trace elements** are of great importance to the growth and development of plants, even though only a minute quantity needs to be absorbed. Nearly all plants receive mineral nutrients in solution through the membranes of the root hairs. Therefore it is not sufficient for the elements required by plants merely to be present in the soil; they must be in a form which can readily be absorbed, and the exchange of ions between the soil particles and root hairs is of immense importance to the plants.

7.3 Organic matter and its decomposition

The remains and excreta from plants and animals are important soil constituents. So are many living organisms. Earthworms, springtails, mites, and insects consume and spread dead organic tissue and mix soil particles. Fungi and bacteria, especially, help to decompose organic matter into water, carbon dioxide, organic acids, and salts, such as nitrates.

Certain bacteria which occur in nodules on the roots of leguminous plants assimilate gaseous nitrogen and convert it into protein, which is broken down by other bacteria into ammonia and nitrates. Nitrogen thus becomes available for plants able to absorb its soluble salts; this is a vital cycle, for although green plants, using solar energy, convert atmospheric water vapour and carbon dioxide into sugars, cellulose and other structural matter, they cannot acquire nitrogen directly from the air.

As roots rot, they leave tiny spaces through which soil water may pass; also, as roots, stems, and leaves decompose, they produce **humus** (a name sometimes misapplied to the surface litter of slowly rotting leaf-mould). Humus is a complex colloidal mixture of substances, black in colour. A colloid contains very fine material, and can either exist as a mobile fluid or, at lower temperatures, as a gel, or flexible solid. Thus it may be present in soil as minute particles or as a jelly coating the mineral grains. About half of it is humic acid combined with various bases. The basic content may be high enough to give an alkaline rather than an acid reaction. This is mild humus or **mull**, as opposed to raw humus or acidic **mor**. Humus plays a very important role in helping the soil retain elements which the plants may subsequently use; especially when combined with tiny electrically-charged clay particles, as **clay-humus**.

The gummy quality of humus is also important in helping to bind large earthy lumps; which makes for soil stability and combats soil erosion. To some extent the attraction between electrically-charged clay-humus particles also helps to build up smaller aggregates.

Soil spaces provide small gaseous pockets essential for the micro-organisms. The bacteria require oxygen. Waterlogging may interfere with this; and cold and drought also limit bacterial activities. In such cases dead organic matter may remain unaltered and peaty in nature; and humus formation is checked.

So before looking at nutrient exchange in more detail, we first see how soils are structured.

Fig. 7.2 The present dark organic horizon, in porous material ejected from Mt Ngauruhoe in New Zealand, overlies a darker layer similarly formed at the surface of tephra from previous eruptions.

7.4 Soil texture

Size of particles

Particle size affects soil properties in a number of ways. Water drains rapidly through sandy soils, but is held in soils of very fine texture. In a clay soil the particles are too small to allow adequate drainage. In some fine soils the water adheres so firmly to the particles as to restrict its availability to plants.

Material larger than 2 mm diameter is classed as **gravel** or **stones**. Below this size are **coarse sands** down to 0.2 mm, **fine sand** to 0.02 mm, and **silt** to 0.002 mm diameter.

The sum of the surface areas of the vast number of minute particles which make up clay is so great that it enormously increases its power of adhesion to other substances. Because of this, elements which plants may need are not so easily washed out. We have already noted the valuable retentive properties of the colloidal **clay-humus complex**.

As Fig. 7.3 shows, soils are broadly described as **sand** if they include about 80 per cent sand to 20 per cent less silt or clay. A **sandy-loam** contains

SOIL TEXTURE

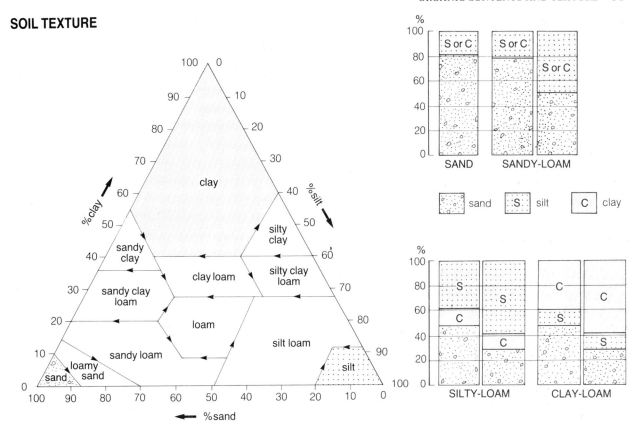

Fig. 7.3   Soil composition in terms of particle size.

between 30–50 per cent sand, and the rest silt or clay. Other types of loam include 30–50 per cent sand, with 30–50 per cent silt, and up to 20 per cent clay. If silt predominates it is a **silty loam**; if clay, then a **clay loam**.

The soil texture can be estimated by sieving it through decreasing meshes, to separate grain sizes; or by comparing the settling rates of particles in water, the rates being proportional to the diameters of the particles.

Grain groupings

The mineral particles tend to bind together as aggregates, or **peds**. The soils may be regularly **granular**; have a crumb structure (**crumby**); be divided into vertical columns or long prisms (**prismatic**); be made up of large, sharp-edged irregular blocks (**blocky**); or have flat, thin horizontal layers (**platy**).

Fig. 7.4   Grain groupings.

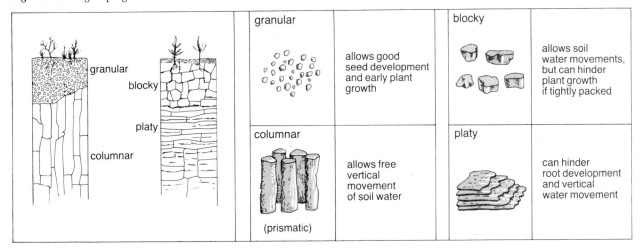

Hard pans and salt layers

Soil textures tend to vary with depth and may change abruptly, as when certain compounds have formed a hard layer. Thus in some podzols (p.91) a layer of iron compounds is deposited not far below the surface. In this case acid rainwater and organic acids, together, dissolve iron oxide (ferric oxide) which is normally insoluble in water. The solution washes down through the soil. But where the soil acidity falls below a certain value, iron compounds are precipitated and an iron pan may build up. In other soils various salts are carried upward in solution and concentrated by evaporation at or near the surface; they may build considerable thicknesses.

Soil colour

Soil colour may, or may not, be significant in indicating the composition of the soil. Very small amounts of a compound may produce an overall coloration no different from that of a soil which contains it in abundance; this is often so with iron compounds. On the other hand colour difference can be significant. Red-brown ferric compounds occur in well-aerated soils, while in waterlogged soils derived from the same parent rock the exclusion of oxygen makes for the presence of bluish-grey ferrous iron compounds.

The fertile chernozem soils of the prairie grasslands (p.94) have a blackish appearance, for humus is well distributed through the profile; though the organic content is usually only of the order of 10 per cent; and indeed the soil may have only half that content and still appear black. Here the dark colour indicates fertility; but this is not always the case. The blackness of volcanic soils of the Indian peninsula is partly due to the presence of titanium compounds. Again, black peaty soils may be organic in nature, yet in this condition infertile. Caution is needed in drawing inferences from soil coloration.

The effects of agriculture

Continual cultivation is bound to affect soil texture. Ploughing, harrowing and rolling are used to break down aggregates to give a finer, more even surface structure. Ploughing breaks crusts and increases porosity; though this may not always be desirable. In hot countries with seasonal periods of rapid evaporation, the use of peasant hoes may be more effective in maintaining soil structure than deep ploughing, which allows upward movements and loss of soil moisture. Elsewhere, excessive tillage of naturally sandy or silty soils may lead to wind erosion.

The use of heavy machinery on wet soils can over-compact the surface, and even cause clay particles beneath the surface to be re-oriented and form a dense horizontal layer – a **ploughpan** which hinders water percolation, aeration, and root development.

The practice of zero-tillage, seeding grains into narrow holes or slits, without turning the soil, is sometimes used where there is danger of deterioration through ploughing. But this usually calls for extra treatment to control weeds and pests, and there is some danger of surface compacting if the land is not ploughed at all.

Farmers often attempt to alter soil texture through addition. In areas of sandy soil a calcareous, silty clay (marl) may be used to increase the soil's clay content and basicity.

7.5   Soil acidity

Acidity has an important bearing on the chemical composition of soils. It also not only hinders the absorption of nutrients by plant roots, but affects the rate of bacterial decomposition. Soils may become acidic because the organic matter which decomposes contains few bases; or because few bases are supplied by the parent material; or because inorganic or organic acids replace and remove metal ions from soil particles. We may regard ions as an atom or group of atoms carrying an electric charge; metal **cations**, have positive charges and form **basic** oxides while non-metal anions have negative ones and form **acidic** oxides. When salts dissolve in water positive and negative ions separate. Both acids and bases also dissociate in solution to give ions. It is possible to indicate soil acidity by measuring **the concentration of hydrogen ions**, provided by the acids. The symbol pH is used to show this concentration.

An **acid soil** is represented by a *smaller* pH value than a basic one, because the *logarithm of the reciprocal* of the hydrogen ion concentration in thousandths is used. Thus 1/10 000 part by weight = pH1; 1/100 000 part = pH2; and 1/10 000 000 000 part = pH7. Hence pH5 represents an acidity ten times as great as pH6, and so on.

Fig. 7.5   pH values.

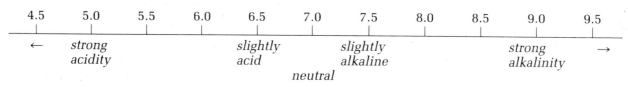

| 4.5 | 5.0 | 5.5 | 6.0 | 6.5 | 7.0 | 7.5 | 8.0 | 8.5 | 9.0 | 9.5 |

←   *strong acidity*   *slightly acid*   *slightly alkaline*   *strong alkalinity*   →

neutral

A soil which is neutral, neither acidic nor basic, has a pH value of 7. A soil with organic acids present, and with salts leached out, so that the clay-humus complex holds hydrogen instead of metallic ions, may have a pH value of 3. A soil containing bases from parent material, such as calcium carbonate, and with the clay-humus complex retaining basic ions, may have a pH value of 9.

If a particular soil is losing bases, and perhaps becoming less fertile on this account, the pH value will tend to decrease. But although the pH values show the balance between the acidic and basic contents, they do not necessarily give an indication of the *quantity* of bases retained in the soil:

7.6   Plant nutrients in soils: removal or retention

Many soluble salts are easily removed as storm water seeps through the soil. Metal ions such as potassium, calcium, magnesium and the ammonium ion, needed by most plants, are particularly likely to be transferred in solution and become unavailable as nutrients. As we have seen, certain soil constituents can check the removal of bases, especially clay minerals in association with humus.

Clay minerals

Clays consist of extremely fine particles, formed by the alteration of various parent materials. They are mainly composed of distinctive clay minerals together with other minute substances.

In clay minerals the atoms of silica, aluminium, and oxygen form lattice structures arranged in layers. The composition of the lattice and the arrangement of the layers characterise an individual clay mineral.

Under various environmental conditions clay minerals change their composition. For instance, under hot, wet conditions they tend to lose silica from their structure, as shown in simplified form in Fig. 7.6 Notice that the layers are indicated, and *not* the internal structure of the molecules. The clay **montmorillonite** is seen to have a 2:1 layered structure, but loses silica as it weathers. Initially, this produces a 1:1 layered **kaolinite**; but further weathering leaves the **aluminate**.

A parent rock such as basalt can thus produce montmorillonite in combination with hydrated iron oxides, then weather to form kaolinite with iron oxides, and finally create alumina with iron oxides, as in bauxite: all the result of **progressive desilification**.

The clays break down through hydrolysis under moist conditions; but the end products, and the retention or loss of these in the soil, vary with the temperature, and with the acidity or basicity of the soil solutions.

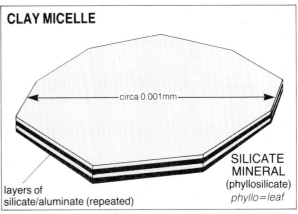

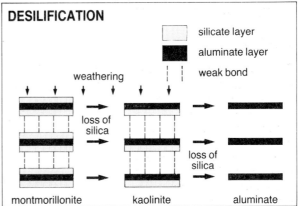

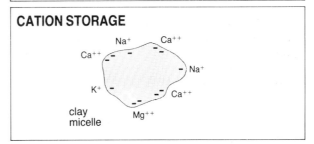

Fig. 7.6 Silica minerals, and progressive desilification.

Fig. 7.6 stresses the large surface area provided by innumerable **clay micelles**. It also shows how the negative surface charges are able to adsorb and retain a variety of cations, and retain these potential nutrients long enough for them to be exchanged and absorbed through the membranes of the root hairs. Fig. 7.7 shows diagrammatically how cation exchange takes place. The hydrogen ions, and also the aluminium ion (Al^{+++}) are acid-generating cations.

Clay–humus particles and base retention

The **colloidal particles** of the clay–humus complex are particularly effective in absorbing and exchanging cations. Their presence helps a soil maintain its natural fertility or hold cations from an added fertiliser. Farmers are particularly interested in the **cation-exchange capacity (CEC)** of a soil. This is a measure of the weight of ions to the weight of soil. The relative magnitude of the units indicates the value of the various soil colloids in making nutrients available and the soil fertile. Thus forms of humus have CEC values from about 200–500; montmorillonite between 80–150; kaolinite as low as 3–15; and the iron and aluminium oxides (left when silicates such as $xFe_2O_3.yAl_2O_3.zSiO_2$ break down) have a CEC of only 4. The montmorillonite has a much greater surface area than the kaolinite produced by desilification.

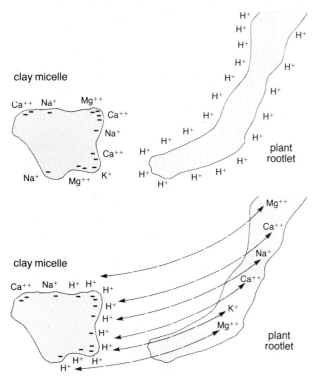

Fig. 7.7 Mineral nutrients become available for plants.

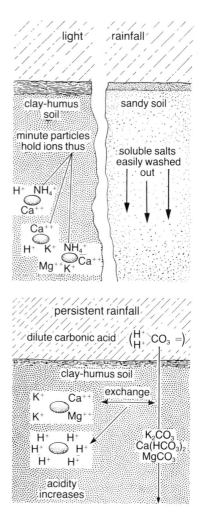

Fig. 7.8 With light rainfall, soils with clay-humus can retain basic ions, in contrast to coarse sandy soils. Heavy leaching under acidic conditions tends to replace and remove the bases. Soluble carbonates and bicarbonates are leached out with other salts.

Under persistently wet conditions percolating acidic water **leaches** out a large proportion of the bases. As Fig. 7.8 shows, rain-water is dilute carbonic acid. When raw humus is present the effect is accentuated. The continued presence of bases in the upper soil must then depend on the ability of the parent rock to maintain a supply through weathering. In soils derived from chalk and other limestones the parent rock is usually able to provide a calcium supply; this helps to maintain the basicity, which suits so many of the grasses grown for grain.

7.7 Climatic influences on soil formation

The effects of excess water passing downward through the soil point to strong relationships between soil formation and climatic influences. In a cool climate **leaching** under acidic conditions leaves silica in the upper soil. But in hot, wet conditions leaching tends to allow silica released from clay minerals to become mobile, so that it leaves concentrations of hydrated oxides of aluminium (bauxite), iron (limonite), and manganese (manganite). Such soils are termed **pedalfers** (Al – aluminium; Fe – iron).

When **drying conditions** persist salt solutions are drawn upward, so that, as evaporation takes place at, or near, the surface, salts are deposited and accumulate. Where dry and rainy periods alternate the salts may accumulate at a depth depending on the balance between upward and downward movements, perhaps a metre beneath the surface, often much less. Alkaline soils of this type, and those with lime accumulations in the form of nodules of calcium carbonate, are known as **pedocals** (Ca – calcium). Some compounds accumulating as layers within the soil are apt to impede drainage, and may lead to waterlogging; iron compounds, in particular, form iron pans which interfere with vertical movements.

The vertical movements are of immense importance to soil properties and to the creation of distinct layers of organic and inorganic compounds at certain depths. The leaching process which removes bases and colloids is known as **eluviation**. The reception of these materials in a lower soil level is **illuviation**. Thus materials, texture, consistency and colour differences may be identified at various depths, down to the parent rock; though some soils are remarkably uniform in these respects throughout their profile.

High temperatures and a ready supply of oxygen favour bacterial activity and speed up chemical processes; so that in warm, moist soils plant decomposition proceeds rapidly. In tropical rainforests there is a frequent supply of plant material to the surface, but its very rapid decomposition provides soluble compounds which are rapidly re-absorbed, or sometimes physically removed. In such a rapid cycle humus does not accumulate in the way one might expect.

High temperatures and strong winds increase evaporation, so that there may be insufficient moisture for shallow rooted plants; or the surface soil may cake or crumble. Winds may also remove nutrients as dry top-soil blows away; though the transport of mixed mineral particles may lead to eventual accumulations of fine fertile loess.

Low temperatures and excess water both reduce the amount of oxygen available in soil spaces, and so hinder bacterial activity. Carbon dioxide is *more* soluble at low than at high temperatures, and although the rates of reaction fall with temperature, chemical weathering by carbonic acid may increase.

The creation of soil particles and the release of mineral nutrients are, of course, dependent on the combined influences of temperature and moisture, and on such processes as expansion–contraction, freeze–thaw, and chemical action.

7.8 The development of soil horizons and nomenclature

The rate at which a soil develops to a recognisably mature form varies considerably with circumstances. **Time** is needed for a balance to be achieved between the numerous physical, chemical and biological processes affecting it. One or more of these processes may become dominant and act rapidly to influence the composition of the soil; but generally a mature balance is achieved slowly, perhaps over hundreds, or even thousands of years. But in the meantime, it will have been affected by erosional processes, by animal burrowing, by surface cracking, by short-term climatic change, and more recently by the increasing activities of the human population.

Recognisable **mature soils**, with distinctive profiles, though not necessarily with clear, contrasting layers, occur beneath undisturbed vegetation formations, such as long-established and largely undisturbed rainforests. These undisturbed areas, sadly, are rapidly shrinking, and soils affected by clearance and cultivation develop their own characteristics. However, we shall look first at certain categories of soils which are related to the larger biomes and thus have a marked zonal distribution (p.88). To illustrate how such soils acquire a recognisable profile we may look at the formation of a podzol (podsol) developed in a moist cool–temperate location on parent material with poor reserves of potential nutrients, and under acid conditions, which mean that there are few microorganisms in the soil.

The podzol, shown in an idealised way in Fig. 7.10, illustrates the effects of eluviation and illuviation, the vertical movements which cause a translocation of minerals through the profile, and lead to the development of layer-like **horizons**.

Characteristic letters are used to describe the horizons, with numbered sub-divisions according to the complexities of the soil and the need for detailed description. In Chapter 8 we consider a number of ways of classifying soils. Some classifications, such as that used here, suit geographers dealing with broad zonal distributions. Others suit those making more detailed local studies, together with soil scientists and farmers considering soil properties which may vary from one field to the next.

In Fig. 7.10 we can see the upper soil in which **eluviation** is taking place, a broad **A horizon**; and a lower zone in which there is **illuviation**, the **B horizon**; beneath, in the **C horizon**, is material weathered from the parent rock, which in this case has poor nutrient reserves.

Fig. 7.9 It is difficult for mature soils to develop in mobile material such as this steep talus.

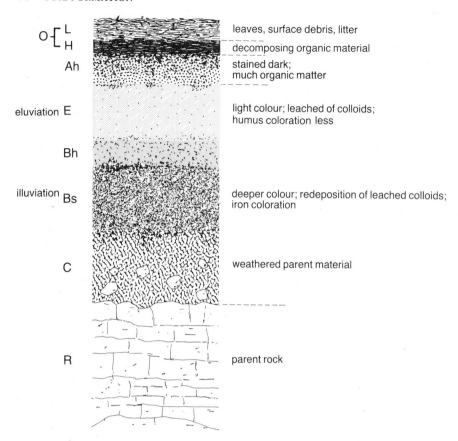

leaves, surface debris, litter

decomposing organic material

stained dark;
much organic matter

light colour; leached of colloids;
humus coloration less

deeper colour; redeposition of leached colloids;
iron coloration

weathered parent material

parent rock

Fig. 7.10   To illustrate the results of weathering, organic decomposition, and the vertical movements of soil solutions, the layer-like horizons are shown diagrammatically. But although horizons may be instantly recognisable in some soils, particularly in the podzols, few are clearly layered in this way. Colours and textures may gradually change, or may be completely mixed by soil organisms. The depth of the soil, and of the horizons, are also very variable. Sometimes the profile is less than 100 mm, sometimes several metres deep.

On the surface there is an accumulation, or **litter L**, of leaves and other fallen plant material; in pine forests the needle leaves are poor in nutrients and decompose slowly. Below this is a shallow horizon where bacteria and fungi are active, and decomposing organic material gives a dark coloured **humus layer H**. In other soils the depth of this layer and its composition depend on the amount and nature of organic matter, on the activity of micro-organisms, and on the rate of removal: so some organic layers may be 50–100 mm thick. The whole of the **upper organic layer** may be designated **O**.

In the A layer there is a regular downward movement of soil solutions. The upper part, sometimes called the A1 layer, is stained with organic matter and dark. This is indicated by the suffix **h**. Beneath, the effects of leaching produce a **lighter zone of eluviation E**, but sometimes shown as an A2 layer.

Some humus-stained colloids begin to accumulate again in the upper part of the B zone of illuviation; this can form a darker coloured, **transitional Bh layer**, often ill-defined. Below this the iron and aluminium sesquioxides (Fe_2O_3 and Al_2O_3) which have been removed and leached down from clay minerals in the E zone, are redeposited where conditions are less acidic, and form a horizon, often stained reddish by the iron compounds, **the Bs layer**.

In other soils various other suffixes are used to identify the materials present and the processes which have caused their translocation (p.90). Fig. 7.10 is merely an indication of the way in which horizons may develop within a soil, and suggests how their presence may allow us to recognise particular soil types.

Remember that vegetation, which is controlled in the main by climate, and the soils are interdependent. Interference with one affects the other; and interference with mature soils produces *new* soil characteristics. Such new soils require new nomenclature which is both description (descriptive) and which deals with the causes of formation (diagnostic) (p.89).

7.9   Soil formation: progessive changes

Using the development of a podzol as an example of the creation of soil horizons may give the impression that podzols form only beneath coniferous vegetation in moist, cool temperate conditions. In fact the formation of a podzolic soil can take place under these climatic conditions on parent rock or regolith deficient in basic nutrients, and we are often able to view the continuous processes in different stages of development.

At first a sandy regolith, or even a sand dune may be covered with a litter which forms an organic-mineral mix, only 100 mm or so thick. As this increases in depth, with decomposed organic matter beneath, a blackish, or dark grey horizon is formed, and the release of sesquioxides causes the underlying sand to show iron staining. At this stage, with **a single developed upper horizon**, the profile is said to be that of **a ranker**.

Then, perhaps over hundreds of years, the iron staining gives way to a lighter eluvial zone, with the translocation of the iron, aluminium and humus colloids to the lower B horizon, with its darker staining.

The vegetation on, say, a sandy dune will almost certainly have varied as the soil formed, the plants forming a succession (p.106) of mosses and lichens, then grasses and herbs, and finally perhaps conifers. Several stages of rankers, podzols and of plant associations may be seen in a single area on dunes of different ages.

With other soils, too, there are **periods of colonisation** of a site by plants and the formation of thin mixtures of minerals and organic matter, before development towards a recognised soil type occurs over periods of hundreds, or thousands of years, depending on relative climatic stability and lack of major interference.

7.10   The effects of relief

Local topographical differences are usually accompanied by local variations in soils and their moisture content, and in the forms of vegetation. Soil creep, or erosion, removes soil from slopes, so that it accumulates at the foot or the floor of valleys. The thin soils of steep slopes, the mixed materials of screes below, and the soils developed on the gentler outwash slopes beyond have different moisture contents, support different plant communities, and acquire different organic contents.

The re-sorting of mineral nutrients may in some cases improve soil fertility. Where the fertility of plateau soils has fallen with time, those developed

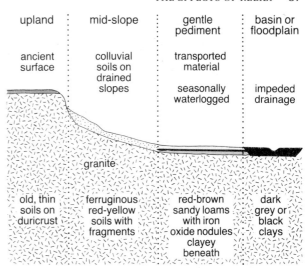

Fig. 7.11 A catena formed under alternating wet and dry seasonal conditions in the tropics. A sequence which occurs in many parts of the East African tablelands. Here soil types are identified, not soil profiles.

on material derived from the erosion of the scarp slopes of the plateau edge are often much more fertile, even though derived from the same parent rock and from the old soils of the uplands.

The term **catena** is used to describe a regular repetition of soil sequences down a slope, which is directly or indirectly due to topography or to changes in parent material. Fig. 7.11 shows a sequence of soil types which is repeated in many parts of the tablelands of East Africa under broadly similar climatic conditions. In practice, the horizontal boundaries are seldom clear-cut; but the typical sequence of this catena can readily be identified where similar topographical elements occur. Sometimes the term 'catena' is used to indicate a characteristic sequence of different soils from the same parent material, but which vary with relief and drainage.

CHAPTER

8

SOIL GROUPS AND
THEIR DISTRIBUTION

8.1 Classification: simple and sophisticated

In the previous chapter we saw that various natural processes combine to create a recognisable soil type. These processes are affected by local temperatures, precipitation, rates of evaporation, and seasonal climatic variations; all of which act directly on soil material and soil organisms. It is not surprising, therefore, that there are broad zones on the earth's surface where reasonably similar climatic conditions aid the development of similar types of soil.

However, in soil formation climate is not necessarily *more* important than the parent material, which in any case affects soil-forming processes. It takes a very long time for a soil to acquire a mature form; hence the climate under which a mature soil has developed may not be exactly that of the present local climate; and, as we have seen, soils may be observed in a single location in various stages of formation. Even a recognisable soil type may be undergoing modifications; not only by climatic elements, but by physical actions, such as erosion, deposition, or changes in drainage conditions.

Thus, as with climates, the concept of zonal soils is disturbed by numerous exceptions. Soils with recognisably similar characteristics occur in a number of different zones – they are **intra-zonal**. These include soils which have developed under waterlogged conditions; soils with a high saline content; and soils formed on the same type of rock, whose main characteristics are due to the nature of the parent material rather than the climate (as in many limestone areas). Some soils are of recent accumulation and are thus immature; like those on alluvial flats, screes, dunes, or lava flows, which have had little time to weather or acquire characteristics due primarily to the climate. Such soils are **azonal**.

Zonal influences have strongly influenced a number of **schemes of classification** which have proved useful to geographers and pedologists, such as that of the American **C.F. Marbut (1928)**, who was influenced by earlier Russian soil schemes of classification. Variations were published in a number of countries and adopted by the US Soil Survey and the Soil Survey of England and Wales. It was a **simple genetic classification**, describing soil-forming environments, and recognising the effects of leaching, drainage and humus formation. In its major categories it separated the pedalfers and pedocals, and pointed to lesser categories of each type as formed under tropical and temperate conditions. Thus the pedalfers included podzols of the temperate humid zone; the ferrallitic soils, rich with iron compounds, of the humid tropics; and plinthites (laterites) with hardened iron pans of the seasonally wet-and-dry tropics. Within this classification there were great soil groups characteristic mainly of vegetation formations which reflected the climate, such as the black earth of the temperate grasslands, known by its Russian name – chernozem. Lower categories distinguished soil types based on surface texture, stoniness, drainage, and responses to the local environment.

However, with the increasingly detailed study of soils worldwide, it was realised not only that there are numerous soils, particularly within the tropics, which do not fit easily into such earlier classifications, but also that many soils owe their characteristics to human activities. It was therefore decided to place soils into categories based on description of their morphology and composition.

The **US Department of Agriculture Soil Survey's 7th Approximation**, a seventh attempt to formulate a classification of this type, has been widely adopted and named the **Comprehensive Soil Classification System (CSCS)**. Ten major orders are recognised, with six lesser categories below. Together they contain almost 20 000 classes of soil, most of which require expert laboratory identification. Nevertheless, as a simple introduction, it is valuable to recognise the descriptions and meaning of the terms used to describe the ten major soil orders, and to understand the way in which the soil horizons are distinguished.

There is as yet no internationally agreed system of soil classification and horizon nomenclature. A **Food and Agriculture Organisation (FAO) system** has also been widely adopted, and a map of soil units prepared by FAO–Unesco. We shall therefore look briefly at the ways in which these two systems communicate information at a level which will be valuable for geographical studies.

8.2 The 7th Approximation's soil orders and horizons

The ten **orders** are as follows:

Entisols: young soils without horizons, as on recent alluvium or dunes.
Inceptisols: moderately developed young soils, where leaching is active, but horizons are only beginning to develop.
Oxisols: old, highly weathered soils of the humid tropics, with few primary minerals (low CEC), and with clays containing the sesquioxides of iron and aluminium and kaolinite.
Ultisols: soils of the warmer mid-latitudes; deeply weathered, with an eluvial A horizon and clayey deposits in the B horizon; low in bases.
Vertisols: tropical and sub-tropical clayey soils with seasonal deep cracks which open and close, allowing surface material to be transferred into their mass.
Alfisols: soils of moderately humid climates, with good base content, but a B clayey horizon created by downward migrations.
Spodosols: virtually podzols; with a pale A horizon (albic) and darker B horizon with sesquioxides (spodic); a low CEC and lacking carbonate minerals.
Mollisols: with many bases, especially calcium, and a soft dark A horizon, with decomposed organic matter (L. *mollis* – soft); found in sub-humid regions, such as the temperate grasslands.
Aridisols: dry soils, with light surface horizons, low in organic material; often with accumulations of carbonates and other salts.

All but the first two take special account of the nature of the horizons. The properties of a **soil horizon** differ from those of the layer above or below it. The characteristics, which can be observed or measured in the field, such as structure, texture and colour are generally given names derived from Latin or Greek, to aid universal use, as we have seen with *mollis* (soft) describing a **mollic horizon**.

There are also sub-orders whose descriptive words contain an element which denotes the soil conditions, such as **xer** (Gk. *xeros* – dry). Thus **xeralf** describes a dry alfisol. Fig. 8.3 shows more clearly the use of suborders.

Then there are descriptive words for particular horizons. Apart from **albic**, **spodic**, and **oxic**, already used above, **argillic** (L. *argilla* – white clay) describes an illuvial horizon where clay minerals accumulate; **salic** a horizon enriched by soluble salts; and **histic** (Gk. *histos* – tissue) a peat horizon. Other names will become familiar by use in their proper context.

8.3 The FAO system and FAO–Unesco soil map categories

In general this uses as many traditional names as possible: such soil units as podzols and chernozems, and others described above, such as vertisols and histosols. Its new words also make use of an adjectival prefix as a means of description: thus **acric** (Gk. *akros* – ultimate) in **acrisols** indicates very strong weathering and low base content, as found in the red–yellow podzols.

The combination of letters used to describe horizons is a clear one, and is generally followed in this book.

Main horizons

L, H, and O: the litter and organic accumulations, as shown in Fig. 7.10.
A   a mineral horizon at, or adjacent to, the surface; usually with humified organic matter mixed with the minerals.
E   a pale horizon with sands and silts of resistant minerals, having lost clays and sesquioxides through leaching; and so an eluvial horizon.
B   generally lacking parent material, but may contain illuvial concentrations of clay minerals, sesquioxides or humus. The material itself may create very different structures – granular, prismatic, or hard pans – and so a suffix is almost always used to give a clearer description: Bh, with humus; Bs with sesquioxides; and so on.
C   a layer of unconsolidated material overlying the rock (R). There may be accumulations of carbonates or other salts. When very compacted, the suffix m qualifies it. Soil horizons which grade one into the other are shown as transitional by using two letters, eg EB, the

first letter being that which the transitional zone most resembles.

Qualifying suffixes

a   an ash-coloured alluvial horizon.
b   a browner eluvial horizon, with some unaltered clay.
c   an accumulation of concretions (eg Bck, illuvial carbonates).
fe   an iron pan.
g   a mottling caused by some oxidation, some reduction, as in gleys.
h   organic matter in an undisturbed mineral horizon; humus content.
k   accumulation of calcium carbonate.
m   strongly cemented (eg mk – by calcium carbonate).
n   accumulation of sodium salt.
p   disturbed by ploughing (tillage).
q   accumulation of silica (eg mq, a silcrete layer).
r   strong reduction – as when waterlogged.
s   an accumulation of aluminium and iron sesquioxides.
t   an illuvial accumulation of clay.
w   change within the horizon reflected by structure and/or colour.
y   an accumulation of gypsum.

8.3   A geographical survey of world soils

The system described above enables the description of tens of thousands of minor soil types, and is of great value to pedologists (who study soils) and for regional and local studies. However, geographers also need to view soils in relation to climatic and vegetational variations on a continental and broad regional basis, even though there are innumerable deviations due to natural processes and human interference.

The sections which follow look at the characteristics of soils in latitudes from the poles to the equator. The names used are both the more familiar zonal descriptive ones used in modifications of Marbut's scheme and the orders, and occasionally sub-orders, of the CSCS system. The nature of the horizons, or the lack of them, is described by the use of letters generally following the FAO system; and the likely acidity is indicated by pH values.

8.4   High latitude gleysols and histosols

Soils of the tundra and peat basins

Under climatic conditions of the ET type described on p.73, the sub-soil remains permanently frozen. Precipitation is moderate to slight, but snowfalls accumulate during winter. Most precipitation follows the spring thaw, so that much of the surface and top-soil becomes waterlogged, and the soil remains moist during summer.

Bacterial action is thus restricted; so under the low, but often close vegetation of tiny shrubs, grasses, mosses and lichens there is a black mass of slowly decomposing plant matter, producing much acid humus. Beneath is a clayey mud, stained blue-grey by ferrous iron compounds, which have not been oxidised to the familiar red-brown of ferric iron. This is known as a **gleysol**, and is formed under bog conditions in other zones.

As winter sets in, the water freezes in the upper layers; this leads to a general increase in volume, with a heaving effect due to the expansion within the soil. The thaw causes a contraction, with quite extensive shrinkage. Stones held in the temporarily frozen zone may thus be stranded on the surface, and conglomerations of stones, some quite large, are a typical feature of the uneven surface. Movements by gravity, after upheaval, tend to sort the stones into irregular rings or polygons. The upward heaving is usually emphasised when the autumn freeze sets in, both at the surface and upward from the permafrost, so that there is saturated stony soil between two freezing layers.

Tundra soils thus contain numerous angular rock fragments, weathered from the parent rock; and, of these, some quite large rocky chunks may move through the soil, impelled by the seasonal freeze–thaw. Owing to this vertical mixing, the soils do not show the strata that characterise acid soils developed under heath and coniferous forest in the more temperate zones.

Fig. 8.1   Gleysols are common in the badly drained tundras; but not all tundra soils are gleysols. Podzolic soils form in the better drained locations.

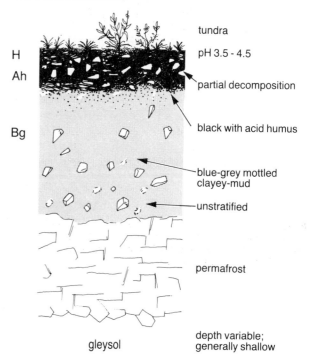

tundra

pH 3.5 - 4.5

partial decomposition

black with acid humus

blue-grey mottled clayey-mud

unstratified

permafrost

gleysol

depth variable; generally shallow

Plate 1 Shallow-rooted coconuts grow in tropical red-yellow soil developed on gneiss in south-west Sri Lanka. A thick whitish zone of kaolin clay with quartz grains is formed beneath.

Plate 2 Terra rossa accumulations provide fertile soil for grains and horticulture near Asteromeritis in central Cyprus.

Plate 3 A deep chernozem profile, rich in humus. The upward movements of bases maintain its fertility, and carbonate nodules appear at the base of the upper profile, where there are no distinct horizons.

Plate 4 A shallow oxisol forms the upper part of the deep regolith formed on weathered granite in eastern Brazil. There is rapid nutrient cycling with the rainforest vegetation above.

Plate 5 Lupins dispersed from small settlements in the Southern Alps of New Zealand have established themselves on river terraces – conifers too – and they line the banks among the dense native forests of evergreen beech.

Human settlement can introduce alien species into undisturbed ecosystems (Plate 5), or destroy indigenous species when clearing for various forms of land use. If protected, the vegetation may regenerate (as in Plate 6), but the original ecological balance is never restored.

Plate 6 Fencing on a New Zealand hillside divides a sub-climax shrub vegetation from cleared, dangerously over-grazed land. The components and balance of the original vegetation can never be restored.

Not all tundra soils are bog soils. Some of those which are better drained do develop stratification of the podzol type described below. They are usually **peaty histosols** bearing a heath vegetation. Some parent rocks also exert their own influence, even here. Limestones, for instance, are readily dissolved by organic acids and carbonic acid, and carbon dioxide is more soluble in water at low temperatures. These alkaline solutions provide bases for the soils, so that the vegetation also responds, and the peaty soil typical of the region may not form. Nevertheless, most of the tundra contains a great deal of ground that is either waterlogged or frozen during the course of the year, and gley soils are common.

Histosols with peat and other organic debris can be found in lowland and upland sites as far apart as Irish bogland, the East Anglian fens and lowlands in Indonesia, and on highlands in Scotland and the Andean countries. The properties and composition of the organic content are influenced by the former plants at the site, and also by the fact that some depressions are likely to receive alkaline solutions, increasing the basicity of the soil material.

As organic soils dry out and are oxidised, nutrients are released; so that well-drained soils may be very fertile, though usually with deficiencies of copper, cobalt and magnesium. **Tropical gleysols** in depressions in humid areas are often used for rice growing, and develop iron-stained horizons under the upper clayey pan created by ploughing – the plough pan.

True podzols (spodosols)

These have been described in an idealised form on p.85. The **spodosols** which are freely drained are often termed true podzols, and described as **orthods** (Gk. *orthos* – true). They occur where there is precipitation through the year, but where lengthy cool, or cold winters check organic decomposition (Fig. 5.20); so that plant matter, partly broken down, remains near the surface, unchanged for many months.

They may develop over large areas where the parent material has a low nutrient content, or on smaller locations, such as fixed dunes. They occur beneath the extensive stands of conifers of the boreal forests, where the trees take few minerals from the soil and their debris is deficient in bases. The humus formed from the carpet of needle leaves is usually strongly acidic.

Even in continental locations where the annual precipitation is small, spring melt may suddenly cause unusually strong leaching. The organic acids remove soluble bases and the sesquioxides from the A horizon, leaving a high silica content. This upper zone of sandy soil is stained in the upper Ah level, but becomes ashy grey in the eluviated E horizon. The darker B horizon, with

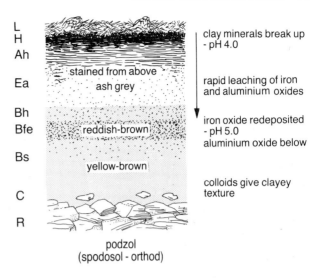

Fig. 8.2 Notice the acidity of the organic layers.

aluminium and iron oxides, tends to have a clayey consistency, due to the accumulating colloids. Sometimes the oxides produce a pan sufficiently well developed to hold up water. At times this gives rise to waterlogging and peaty conditions in the upper soil.

Owing to the lack of bases in the A horizon, lime and other fertilisers are generally necessary before such soils can bear the types of crops which will flourish in these climates.

In the true podzols the horizons are often clearly visible. There are few mixing agents, and the lack of earthworms, especially, helps to maintain the sharp divisions. But not all podzols have such clear profiles. Where trees other than conifers occur under somewhat milder conditions, and the humus is less acidic, and leaching less effective, the horizons become blurred. The profile itself can vary from 100 mm or so to over a metre in depth.

8.5 Mid-latitude soils: from pedalfers to pedocals

Here there are considerable variations in soils corresponding to maritime and continental conditions and differences in latitude. Soil characteristics also respond to the corresponding forms of vegetation and their ability to protect the surface and provide organic material.

The close relationships and interactions between the plant cover and the soil can be illustrated by a westward journey from the maritime north-east of the USA (Fig. 8.3). The annual precipitation decreases towards the semi-arid interior, where there is a tendency to a summer maximum and greater temperature ranges. In general, the change in vegetation from dense mixed forest to dry grassland and semi-arid scrub is mirrored by a succession of recognisable soil groups. Again, one must emphasise that there are countless variations due to local parent materials and drainage, and to larger features like the Appalachian ridges and Nebraskan sandhills. But the soils in the east are mainly pedalfers, and those in the western interior mostly pedocals.

Of course, these soils occur in other continental areas, as Fig. 8.6 shows; so we shall treat each as a representative of a sub-order of one of the major groups (orders) of soils; and remember that elsewhere there are other sub-orders, typical of an environment which is drier, or wetter, or colder, or affected by ploughing.

Grey-brown earths (alfisols)

Afisols occur where annual rainfall is of the order

Fig. 8.3 Illustrating the changes in types of soil from the wetter eastern coastlands to an interior which is much drier and with greater temperature ranges. The soil orders and sub-orders of the CSCS system are used to describe significant features within the areas shown below.

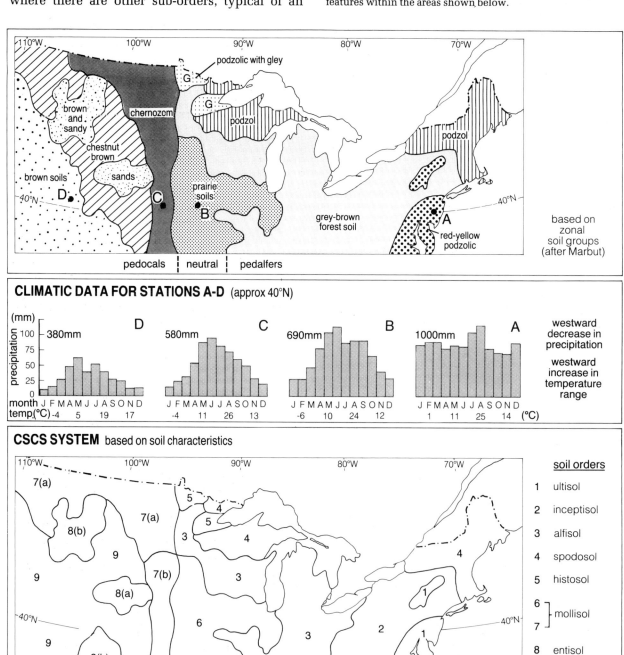

of 750–1000 mm, and where higher temperatures allow the growth of broad-leaved deciduous forests. The thick leaf debris returns many of the bases abstracted by the trees. Bacterial action is also relatively rapid; so although the organic horizon is still acidic, **mild humus** is produced, much richer in plant foods than the **mor** beneath the conifers.

In the northern hemisphere these soils occur not only in north-eastern America, but in western Europe, European Russia, Japan, and north-east China. The forests are by no means entirely deciduous, especially in the northern parts, where stretches of coniferous forest intermingle with broadleaf woodland. There is a considerable variety of flora and soil types, and the parent rock is often a dominant factor in these latitudes. For instance, in north-west Europe porous conditions on outwash gravels and sands allow heaths and conifers to develop at the expense of other forms of vegetation, and podzolisation is rapid and more complete. But grey-brown earths are widespread within the regions mentioned, and are used for rotation farming on a scale which has supported dense populations over the centuries, so that Ap is often a more appropriate description of the upper horizon.

As the soils are less acidic and less leached, so the Ah horizon is a browner colour than that of the true podzols; but it tends to become grey-brown with depth, for the break-down of clay minerals does occur. Colloids with bases and iron and aluminium sesquioxides are carried down, so that

the upper parts of the B horizon becomes darker again.

Soil fauna are more plentiful under these conditions, so that earthworms and other organisms redistribute the soil contents, and make the horizons less sharp. Tree roots often penetrate the bed-rock, or at least reach the sub-soil, and bring up bases, which leaf-fall eventually returns once more to the soil.

Transitional mollisols: the 'prairie' soil

The **mollisols** have a dark brown to black upper horizon, more than 100 mm thick, and of a loose, soft structure when dry. The natural conditions under which they have formed seems to have favoured grassland rather than forest, but with sufficient moisture to warrant the **udoll** description (Fig. 8.3). In the USA they occur westward of the forest earths, where the annual precipitation falls to about 600 mm, and in eastern Europe are found to the east of the grey-brown earths. In each case the original vegetation and the soils themselves have been much disturbed by clearance and cultivation.

Precipitation, especially during the heavy summer storms, is sufficient to carry calcium carbonate and some of the soluble salts deep into the soil. But the base-holding clay minerals are little affected, and **the cation exchange capacity (CEC) is high**. The tall grasses of the natural vegetation annually provide much organic matter above and within the soil. Dark humus coloration is typical;

Fig. 8.4 Soils which have supported large numbers of people farming land cleared of temperate deciduous forest.

Fig. 8.5 Soils with a high cation exchange capacity – widely cultivated in the Corn Belt of the USA.

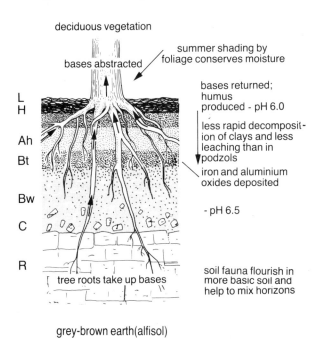

grey-brown earth(alfisol)

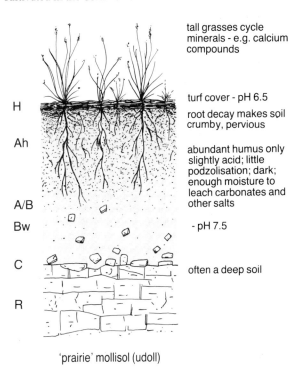

'prairie' mollisol (udoll)

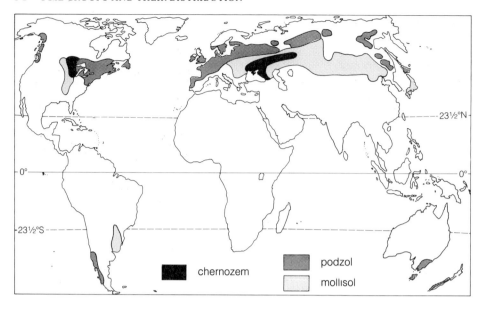

Fig. 8.6 The broad distribution of podzols and mollisols in North America is mirrored in Euro-Asia. In the southern hemisphere mollisols occur beneath the pampas of Argentina and Uruguay.

chernozem

podzol

mollisol

and though decay is slow during the cold winters, decomposition in early summer gives the soil a deep brown colour, finely divided humus, and its crumby structure.

The soils are transitional between pedalfers and pedocals. They have not the calcium carbonate accumulations of the chernozems (Fig. 8.7), nor are they heavily leached of bases from the upper horizons. The A and B horizons are, in fact, difficult to distinguish.

Under fairly humid conditions, these deep, nearly neutral soils, of good structure, have proved particularly suitable for arable farming. In the USA they stretch through the Corn Belt, and further south into Oklahoma and Texas.

Chernozem (boroll)

This black earth **mollisol** belongs to the cold winter prairies and steppes, and hence is described as a **boroll** (Fig. 8.3). In America few of the natural grasses which dominated these ecosystems remain over this great agricultural region. But before being cleared for cultivation they had built up a dense sod cover. Within the soil their extensive root systems decayed, providing humus, distributed through the deep upper soil, and producing a fine crumby structure.

The grasses extract many bases from the soil, especially calcium, and return a litter which produces dark humus. This sometimes fills the burrows of small animals. Worms and other soil fauna also mix the organic and mineral matter.

During the hard winter there is little organic decay; and little leaching until the spring snow-melt. In summer rapid evaporation, helped by wind over the flat landscape, and evapotranspiration from grasses and other herbs, bring up soil solutions. Occasional torrential summer storms interrupt the upward movement and leach down some of the sodium and potassium salts; but these dark soils have **a high pH value**, and under basic conditions the clay minerals do not easily dissociate. Subsequent upward movements cause calcium carbonate nodules to form, perhaps with concretions above the weathering parent material.

Fig. 8.7 The soils which lie beneath large stretches of the central prairies and steppes.

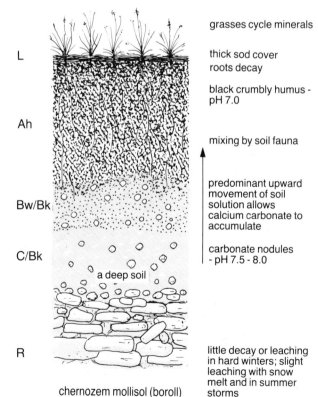

winds relatively uninterrupted by grass cover; strong drying action

grasses cycle minerals

L

thick sod cover
roots decay

black crumbly humus - pH 7.0

Ah

mixing by soil fauna

predominant upward movement of soil solution allows calcium carbonate to accumulate

Bw/Bk

carbonate nodules - pH 7.5 - 8.0

C/Bk

a deep soil

R

little decay or leaching in hard winters; slight leaching with snow melt and in summer storms

chernozem mollisol (boroll)

Mostly the nodules occur about a metre down, but the drier the conditions, the nearer the carbonate-rich horizon to the surface.

The profile is developed upon a variety of parent rocks able to support grassland. Particularly favourable are the unconsolidated calcareous sediments, such as **loess**. The deep, dark homogeneous Ah horizon merges into what can hardly be called a B horizon but yet receives some colloids, and in turn gives way to the light, weathered material beneath.

The chernozems take their name from the black earths of the Russian steppes, where loess is the parent material of much of the moister south-west, to the north of the Black Sea. They stretch away north-eastwards, developed beneath meadow and tussock steppe; but where the grasses become shorter and tussocky the chernozems give way to drier chestnut-brown soils.

These are fertile soils which have given reasonably good yields for long periods when under plough. They retain water well, and ploughing mixes the humus–mineral contents still further. But on the American prairies, and on the steppes, many decades of continuous use has depleted even these huge nutrient reserves, and extensive farming proceeds with widespread fertiliser applications.

Black earths of this kind are found in rather different climatic conditions, under grasslands developed on limestone; sometimes they form extensive enclaves in otherwise forested country, as in the black soil belt of Alabama.

Chestnut-brown and brown soils (aridisols-argids)

On the drier margins of the prairies there is a slower vegetational growth rate. Grasses are shorter, or grow in separate tussocky clumps, and provide less organic matter. The A level is chestnut-brown rather than black. There are strong upward movements of soil solutions after snow-melt, rainy periods, or irrigation. But many have a light yellowish Bw horizon with illuvial clay minerals, received during a wetter period in the past, **an argid horizon** (Fig. 8.3), hence the description. Such a description shows the value of a classification which is able to describe and diagnose horizons, rather than describe only typical zonal soils.

Calcium carbonate nodules accumulate in the B layer, the quantity increasing with aridity, while the depth of the soil decreases. The carbonate tends to give the upper soil a prismatic structure and make it friable.

These soils have supported grasses sufficient for nomadic grazing. At times people have used them for grain growing, but they have rapidly deteriorated. Erosion has been severe in this soil belt in the American west. Dry farming methods have partially succeeded; but irrigation, especially of

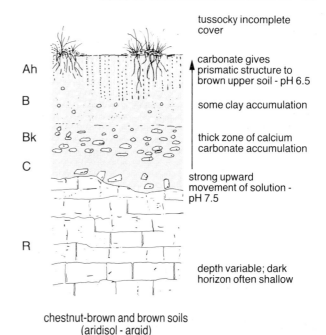

chestnut-brown and brown soils
(aridisol - argid)

Fig. 8.8 These occur mainly between the short grass and the tussock grass country of the drier margins of the prairies and steppes, and have tempted people to cultivate unwisely.

fodder crops, and controlled grazing have been profitably combined.

Under still drier conditions, lighter coloured brown soils have developed under patchy grassland and sage brush, which provide even less humus.

In the inner basins and plateaus of the western USA, and to the south and east of the chestnut-brown soil belt in Euro-Asia, are greyish soils with very little humus and thick deposits of calcium carbonate, or calcium sulphate, close to the surface – **caliche**. These **aridisols** are usually described as **orthids**, or more truly representative of the arid regions. In hot areas the soils are often stained red with iron compounds. In places where the soil has not been leached of gradually accumulated humus, irrigation, with further humus-building, may allow them to be cultivated. Such soils are also termed **sierozems**.

Salinisation

Some arid parts of the world are underlain by highly saline groundwater. Intense evaporation and capillary rise causes salt deposition at or near the surface. Soils may acquire a greyish saline crust (**solonchaks**). But periodically heavy rainfall can leach down the sodium chloride or sulphate and form a highly saline accumulation in the B horizon. The soil above remains structureless and infertile, but sticky when wet and hard when dry, and is traditionally called a **solonetz** soil.

Plate 8 in the colour section shows an example of migrant saline groundwater from irrigation

Fig. 8.9 Winter rainfall in Cyprus readily infiltrates the deep regolith and jointed bedrock of this limestone. Residual wind-blown particles help to form soil material. The low shrubby plants (garrigue) must tolerate both the dry conditions and soils with a high pH value.

areas creating such surface deposition of salt that it destroys soil fertility.

Inland drainage areas may become covered with salt crusts through the contribution of streams which have washed over or through soils with rock salt, for sodium chloride is much more soluble than calcium carbonate.

Much salt, surprisingly, reaches the interior in the form of minute particles formed by evaporation over ocean surfaces. Where they are carried inland in the atmosphere to relatively dry areas they slowly accumulate over long periods.

Dominance of a parent rock: soils formed on limestone

In the temperate regions in particular, limestones have certain characteristics which influence the nature of the soils. Those derived from limestone are basic by nature, and even where podzolisation occurs, in moist locations, have a higher pH value than those on other local rocks.

Calcium carbonate, which makes up their bulk, is soluble in slightly acid rain-water. Thus any insoluble particles which may be present as impurities in the limestone, or are held in the limestone mass, are left as residues when the carbonate dissolves. **Clay-with-flints**, which often lies directly on chalk, is formed in this way, and also the **terra rossa** developed on limestones in Mediterranean regions.

Under warm semi-arid conditions, the carbonate dissolves during wet periods; but with a basic soil the iron oxides tend to remain as insoluble particles high in the profile. As the calcium carbonate is weathered away, this tends to accumulate as a residue with other insoluble minerals, and very slowly produces a layer of red soil. It is

estimated that a loss of some 400 metres of limestone would produce a metre thickness of this terra rossa, so the process is slow indeed. If the vegetation contributes little humus, the iron-red colour is the dominant one.

In temperate conditions which would normally favour podzolisation, and lead to the formation of brown soils under forest and grassland, a parent limestone is often covered by a very dark soil known as **rendzina**. Like many limestone soils it tends to be thin, but has a brown, or black, friable Ah horizon, rich in humus and calcium, covering a grey or yellowish horizon with many limestone fragments.

Fig. 8.10 Particles accumulate in solution hollows in karst limestone in Malta, giving pockets of red soil, which supports small xerophytic plants.

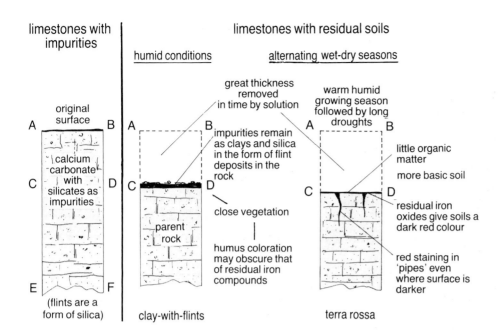

limestones with impurities

limestones with residual soils

humid conditions

alternating wet-dry seasons

original surface

great thickness removed in time by solution

warm humid growing season followed by long droughts

calcium carbonate with silicates as impurities

impurities remain as clays and silica in the form of flint deposits in the rock

little organic matter

more basic soil

close vegetation

parent rock

humus coloration may obscure that of residual iron compounds

residual iron oxides give soils a dark red colour

red staining in 'pipes' even where surface is darker

(flints are a form of silica)

clay-with-flints

terra rossa

Fig. 8.11 Limestones may bear residual soils formed by slow accumulation, as above, together with materials transported from other sources by wind, water, or even ice; but as time passes their ability to absorb water, and their chemically basic nature, strongly affect the soil properties.

8.6 Transitional, sub-tropical soils

In a sense these are not simply sub-tropical, and may be seen, in their northern-most extension in Fig. 8.3. They belong to the great order of **ultisols**, with a clayey illuvial horizon and a fairly low supply of exchangeable bases, found in many parts of the tropics.

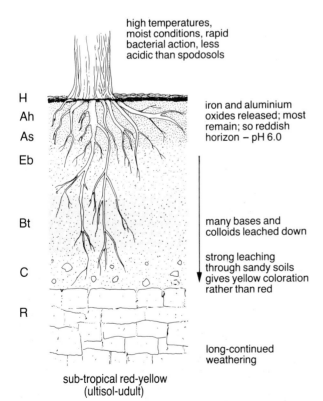

high temperatures, moist conditions, rapid bacterial action, less acidic than spodosols

iron and aluminium oxides released; most remain; so reddish horizon – pH 6.0

many bases and colloids leached down

strong leaching through sandy soils gives yellow coloration rather than red

long-continued weathering

sub-tropical red-yellow (ultisol-udult)

Fig. 8.12

Sub-tropical red-yellow soil – ultisol

These occur in regions with abundant rainfall, but have developed under warmer conditions than the podzols seen in Fig. 8.3. They are found in the warm, moist south-eastern parts of the USA, but occur widely in south-east China and various moist tropical locations.

There is strong leaching as with the podzols. Despite the heavy leaf-fall from a dense forest cover the humus content is low; for bacterial decomposition is rapid during the hot, moist summers and mild winters. Under such conditions the iron and aluminium oxides tend to remain in the upper part of the soil, even though most soluble bases and colloids are washed downward to form a darker Bt horizon.

The remaining iron compounds give the upper parts a typical reddish coloration; though in strongly leached soils, on sandy belts, for instance, the soils are yellow, or greyish rather than red. These are generally of the sub-order **aquults** (wet), rather than the **udults** (moist) described above.

8.7 Tropical soils

Ferrallitic soils (oxisols)

In the humid tropics there is rapid weathering and break-down of the clay minerals, but strong leaching and so **a low CEC**. Plant litter is also quickly decomposed and, in the case of rain-forests, rapidly recycled, so little organic matter accumulates.

As the clay minerals decompose under moist, hot conditions, soil solutions leach down silica as well as bases; some is lost to the soil, but silica is also redeposited low in the profile, and sometimes forms a **silcrete** layer. The oxides of iron and aluminium remain, the iron giving the soil a red or red-brown coloration. In very wet conditions much of the iron is in a hydrated form, so that the soils are yellowish in colour.

On more acid igneous rocks the soils are **kaolinitic**, rather than ferrallitic, and the silica is not completely removed by leaching. Further desilification forms **bauxite**, used as a major source of aluminium.

The soils become deep when formed on a level surface, but even on gentle slopes they are easily eroded away, particularly when protective vegetation is removed. Deep soils formed under a persistently humid climate may develop other characteristics when subjected to different climatic conditions. Thus ferrallitic soils of the tropical wet and dry regions show responses to the downward movements of soil water during the wet season and the upward movements during a prolonged dry season.

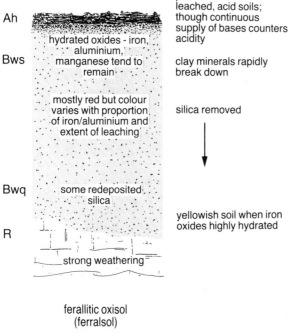

abundant rainfall
soil temperature circa 25°C
soil organisms very active
rapid organic decomposition

Ah

Bws

pH 4.5 - 5.5
leached, acid soils; though continuous supply of bases counters acidity

hydrated oxides - iron, aluminium, manganese tend to remain

clay minerals rapidly break down

mostly red but colour varies with proportion of iron/aluminium and extent of leaching

silica removed

Bwq

some redeposited silica

R

yellowish soil when iron oxides highly hydrated

strong weathering

ferallitic oxisol
(ferralsol)

Fig. 8.13

The soil becomes clayey and plastic during the rainy period, but with alternate wetting and drying the sesquioxides tend to bind together into lumpy concretions in the upper soil. In this form the soil may be freely drained, though root development may be difficult. In some soils the sesquioxides slowly develop a concreted **plinthite** layer, which affects plant development. This is sometimes called a laterite layer (L. *later* – a brick).

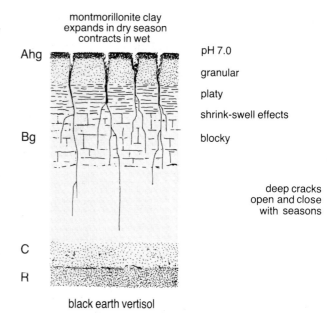

montmorillonite clay
expands in dry season
contracts in wet

Ahg

pH 7.0

granular

platy

shrink-swell effects

Bg

blocky

deep cracks open and close with seasons

C

R

black earth vertisol

Fig. 8.14   The seasonal expansions/contractions allow a gradual vertical re-distribution of minerals. The cracks themselves allow organic material to be engulfed.

Over considerable areas surface erosion has exposed this hard concretion as an infertile surface – a **duricrust**. In some cases these hard crusts occur in dry locations, relics of earlier wetter periods.

Millions of termites help to mix the upper parts of these soils. Large colonies build huge termitariums, mounds of soil rising several metres above the level of the tropical grasslands (Fig. 10.18). They take in and redistribute organic material from the surface.

Fig. 8.15   The development of a plinthite layer.

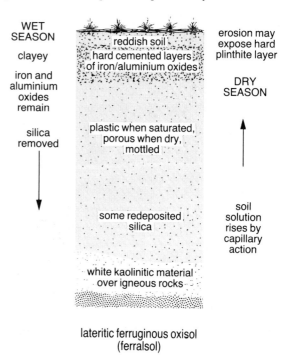

WET SEASON

clayey

iron and aluminium oxides remain

silica removed

reddish soil
hard cemented layers of iron/aluminium oxides

erosion may expose hard plinthite layer

DRY SEASON

plastic when saturated, porous when dry, mottled

some redeposited silica

soil solution rises by capillary action

white kaolinitic material over igneous rocks

lateritic ferruginous oxisol
(ferralsol)

Tropical black earths (vertisols)

Other widespread tropical soils are the **black earths**, deriving their colour from a **montmorillonite** clay complex, though the humus content is low. The high clay content shrinks and swells with changes in soil water storage, and surface-heaving produces irregular mounds and ridges. In the dry season they become deeply cracked. Organic material enters the cracks; and when the rain comes surface soil also slips in and is engulfed. The alternate swellings and contractions cause a slow vertical exchange of materials. The alkaline **vertisols** tend to be very dark; those derived from calcium-rich rocks may develop a horizon of calcium carbonate at the maximum depth of the cracks.

The black cotton soils of the Indian Deccan often have very little humus, and owe their coloration to titanium salts derived from basalt. They are high in exchangeable bases and the soil can absorb water, though much of it is held by the montmorillonite clay and so is not available to plants. The soils have a sticky plastic consistency during the wet season and are difficult to till; but in the dry season the cracked surface tends to crumble when broken by ploughing. But when carefully irrigated and well-managed they prove very fertile.

They also cover large parts of inner eastern Australia, and the Sudan (Fig. 8.16); and occur on flood plains in east Africa and, widely in the Parana river basin in South America.

Other tropical soils

The ferralitic soils and black earths cover huge areas within the tropics; but there are, of course, many other soils developed in particular localities. A freely draining sandstone, poor in mineral bases, allows **tropical podzols** to form. The true rainforest then gives way to scrub-forest, or heath-forest, as in the Guiana Highlands and many parts of south-east Asia.

There are also innumerable **catenas** where undisturbed duricrust soils of plateau surfaces give way to those of free-draining slopes and the pediplain soils beyond, as shown diagrammatically in Fig. 7.11.

8.8 Soil profiles and patterns in small areas

Soil formation in microhabitats

In even more restricted areas, quite small variations of climatic and edaphic (soil) factors can be of great importance to plant life. Near ground level a small ridge or an isolated boulder may have one side directly lit and heated by the sun, the other shaded, cooler, and perhaps moister. One side may be more sheltered from the wind, and there may be a difference of humidity on, and immediately above the surface on either side. In high latitudes, the north and south side of tussocks can have summer temperatures differing by 20 C°.

One side of a wall, or rock, may be noticeably weathering more quickly than the other, and so releasing more potential nutrients. Lichens, mosses or ferns may be able to establish themselves in a thin, nutrient-holding soil layer. If the surface is able to retain their organic debris, higher plant forms may begin to take over, even at this micro-level.

Fig. 8.16 Oxisols cover large areas of the humid tropics, but vertisols are formed where seasonal shrinkage and cracks in clayey material allow vertical exchanges through the soil profile.

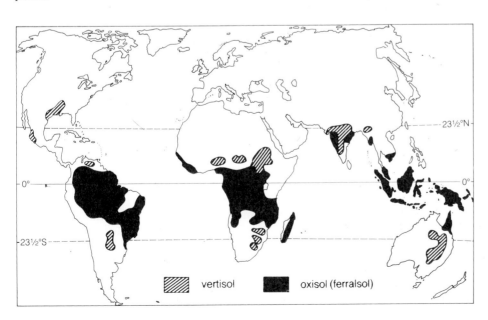

vertisol   oxisol (ferralsol)

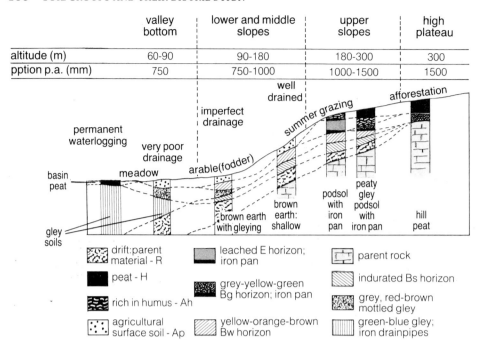

	valley bottom	lower and middle slopes	upper slopes	high plateau
altitude (m)	60-90	90-180	180-300	300
pption p.a. (mm)	750	750-1000	1000-1500	1500

Fig. 8.17 A section showing soil profiles formed under moist conditions in western Britain, from a high moorland to the adjacent valley floor. (After Taylor[2].)

After considering soils on a worldwide basis, it is essential to realise that soil processes are seldom constant, and that particular soil profiles seldom dominate over a wide area. Climatic changes, the removal or deposition of surface materials, and accompanying changes in drainage and soil water content, suggest that the development of the profiles we observe today has probably taken place through periods with considerable environmental variations. This applies particularly to the northern cool temperate lands, affected by alternating glacial and warm periods; so that soils in the British Isles, for example, may have begun to develop under periglacial conditions, perhaps on glacial drift, then under increasingly warmer conditions, until several thousand years ago somewhat cooler conditions set in. These temperature changes were accompanied by long- and short-period fluctuations in precipitation.

Much of this is revealed, for example, by a sequence of soils on upland slopes and adjacent lowland in western Britain, Fig. 8.17. It also emphasises that soil properties reflect people's impact on the landscape in the past and at present.

Soil profiles from moorland to valley

Fig. 8.17 shows a high moorland, with soils formed on parent material derived from the resistant rocks of the upland. But remember that most of Britain's moorlands bear secondary vegetation in response to the changes brought about by the clearance of the former tree cover. Today, cloud and rain for much of the year cause an annual excess of precipitation over evaporation. On the moor the growing period is a short one.

As the wet moorland plants slowly rot down, acidic organic matter accumulates, so that **hill peat** is formed. Beneath this is a **greyish-green gley horizon**, saturated with water held above an iron pan which has developed within the B layer. But the rest of the countryside retains much of its covering of glacial drift, which is thin on the upper slopes and much thicker on the lowland.

As the **slope increases** and drainage improves, the peat gives way to **peaty gley podzols**, with traces of peaty matter and iron pan development in the **profile**.

The **lower and middle slopes** receive somewhat less rain and have the advantage of adequate drainage, better aerated soils, and a rather longer growing season. Shallow **brown earths** on the slope give way to deeper brown earths lower down, on the more gentle, but less well-drained inclines.

In particularly wet locations, the flat, or concave relief of the **adjacent lowlands** leads to waterlogging. The soils are wet and heavy, and the growing season is apt to be reduced by valley frosts. **Gley horizons** are developed again. In valley bottoms which are fully waterlogged, **basin peat** is sometimes formed above thick gleyed layers.

Long-continued human occupation has added its complications. Today part of the upper moor is re-forested with conifers. The valley remains largely meadow land, with the poorly drained parts useless for agriculture. But on the lower and middle slopes ploughing has produced a variety of **agricultural surface soils**, shown as Ap.

SOIL VARIATIONS ON SLOPES AND CLEARED DOWNLAND

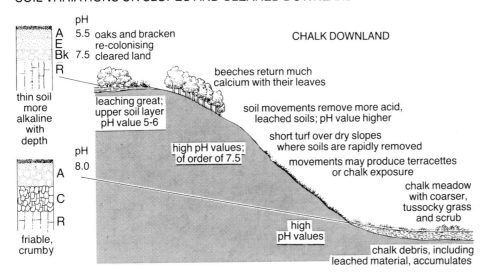

Fig. 8.18 A cross-section of a chalk downland with an indication of the differences in soil profiles and pH values above and below the scarp face.

Chalk downs and lowland in southern England

Interactions between soils and vegetation produce remarkably consistent **catenas** on the chalk downlands, escarpments and adjacent lowlands of southern England.

Leaching is at its strongest on the downs themselves and on the summits above the steep scarp slopes. This removes calcium from the surface layers, so that the pH of the top-soil is lower than that of the soil beneath; alkalinity increases with depth.

This layering occurs even in shallow soils. Many species of plants have a narrow range of tolerance of acidity/alkalinity; so that very shallow acid-tolerant species, such as ling and other members of the heath community, can be found growing side by side with deeper-rooted plants which thrive on basic calcareous soils – calcicoles.

Beech trees tend to check developing acidity by returning calcium in their leaf litter. But where high chalk country has been cleared of vegetation and leaching is strong, heath moor rather than beech tends to take over land which has been abandoned. Under these conditions oakwood and bracken are often the dominants in the vegetation.

Fig. 8.18 shows, diagrammatically, a typical **downland-scarp-valley profile**. The oak and bracken vegetation is established at the summit where pH values are low. On the shoulder, where the downland begins to drop, sufficient chalky debris can accumulate to allow a restricted area of beech wood, with trees rooted in chalk – a hanger on the upper slopes.

The eroded valley sides are not leached in the same way. In places the slope may be too great for soil formation; otherwise a short turf on soil with a high pH value provides dry, chalky pasture.

Debris from the leached material and chalk is washed down and accumulates at the foot of the downs. Here chalk meadows, with coarser, tussocky grass and scrub is likely to be found.

Again, the actual vegetation depends on the nature and extent of the past and present land use. Since early clearance removed higher plant forms, grazing has tended to prevent the regeneration of small trees and shrubs on the downs and slopes. Nevertheless, much chalk country bears characteristic shrubs, scattered over the less closely grazed hillsides – dogwood, hawthorn, roses and the climbing Travellers' Joy. There may also be other plant associations suited to superficial deposits such as clay-with-flints – itself a product of protracted processes of leaching (p.96).

These two simple profiles show that in quite a small area, even on one type of parent rock, there are often significant soil differences, reflected by the various forms of vegetation; and also that, in time, new parent material may be introduced – glacial drift, talus debris, or alluvium – and cause soil processes to develop a different sequence of horizons; perhaps, via rankers, to produce a new mature soil profile.

9

PLANTS AND THEIR ENVIRONMENT

Before considering the composition of the great biomes and the cycling of matter and energy in these complex systems, we need to appreciate the **mechanisms of plant growth**, and look at **the adaptations** which allow plants to compete for nutrients, energy and space, and so create a particular form of vegetation.

9.1   The plant

Plant organs

The higher flowering plants, in the evolutionary sense, have certain characteristics in common. Roots help to hold the plant in the soil, and absorb water with soluble nutrients. Stems hold up the leaves and reproductive parts. The leaves absorb the carbon dioxide and water vapour which, with the help of solar energy, are converted to carbohydrates. They take in oxygen for plant respiration and so release energy for plant processes. In most plants flowers develop seeds for reproduction.

Plant organs increase in length as their cells divide to form more cells at the growing points, near the tips of the stem and roots. Various stimuli, such as light energy, act to affect the process of cell division, causing plant growth to take place in a particular direction, and so cause the organs to bend. Other circumstances such as the availability of moisture and oxygen, determine whether branching takes place or not.

Cells in the plant have shapes and structures appropriate to their various functions. Some develop in ways which enable a plant to survive and flourish in a particular environment: by providing, for instance, a thick bark or a waxy surface which guards against excessive loss of water by transpiration.

9.2   Responses to environment

Plants tend to thrive only where physical and biotic conditions, and competition with other plants, suit their particular species. One, or more, of the many environmental elements which act on plants, directly or indirectly, may determine whether a species may survive or not. The interrelation of these elements is often crucial. Plants need water to survive: there may be sufficient in the soil, but it may not be available when the temperature is too low for them to absorb it, or because it is retained by a particular soil complex (p.80).

Water: shortage, abundance and availability

The rate at which solutions are absorbed through the roots, their distribution through the plant organs, and the passage of water through leaf pores (stomata) are all affected by the temperature and humidity of the surrounding air. Different plant organs have different responses to water shortage or abundance; for some are within the soil and others are perhaps several metres above it.

Plant growth and survival are affected by the nature, amount, and seasonal distribution of precipitation. Seasonal abundance followed by lengthy drought favours plants with mechanisms enabling them to conserve water. While the same amount of rain received at short intervals through the year will allow quite different species to become dominant. Precipitation as snow may blanket the ground and protect plants against excessive cold, and also provide a store of water for release during the spring melt.

The water-retaining properties of soils obviously affect the vegetation, but so does the nature of the surface. Torrential rain that runs off rapidly does not even enter the soil storage, and so is unavailable to plants. Water held at the surface may be lost when conditions favour rapid evaporation.

Excess water, leading to aquatic conditions or waterlogging, obviously favours those plants able to receive sufficient aeration. The water hyacinth, for instance, floats freely, bouyed up by its bulbous leaf stalks. Plants which flourish in a watery environment are termed **hydrophytes**. Those tolerating wet, marshy conditions are **hygrophytes**. Those which live where there is neither water excess nor deficiency are called **mesophytes**; and those with organs adapted to enable them to obtain and conserve water in arid regions are **xerophytes**.

Responses to temperature

For each plant there are maximum and minimum limits for normal growth. There are also limits beyond which the plant cannot survive. Within these limits lies **a range of temperatures most favourable for plant growth**.

A high or low temperature combined with other conditions, such as light intensity or air humidity, particularly affect a given plant. During the exceptionally bitter winter of 1962–63, olive trees in southern Italy survived periods of dry cold; yet, shortly after, a cold damp spell, with no lower temperatures, caused considerable damage to the tissues of leaves and buds.

Different parts of a plant may be harmed by quite different temperatures. Seeds, for instance, are usually more tolerant than the shoots. Also the effects of a given temperature on a plant depends on its state of development, for example whether it is the blossom or the fruit which is exposed to those conditions. Plants live under different temperature conditions at various stages of growth. **The stems, leaves and roots systems develop in different environments**.

The temperature within and close to the surface is especially significant for young plants. A station in Java recorded absolute maximum and minimum temperatures of 35·8°C and 18·3°C in air 1200 mm above the surface, while beneath the surface they were 31·5°C and 23·4°C at 30 mm depth, and 30·1°C and 26·9°C at 900 mm depth. The mean annual temperatures at these levels were 26·1°C; 29·2°C and 29·5°C respectively.

The surface albedo is important, and the colour and texture of the soil affects its temperature. In Bedfordshire, during an afternoon in early summer, thermometers buried 100 mm under loam, loam and soot, and loam and lime recorded 18°C, 22°C, and 15°C respectively.

During daylight hours there are continuous exchanges between the lower air and that above. As the lapse rate in air immediately over a heated surface may be hundreds of times the environmental adiabatic value, there is turbulence. This, of course, influences the rate of evaporation and so affects the plants, even when there is little horizontal movement of air. Conversely, on still nights when the surface rapidly loses heat and chills the air in contact with it, moisture from the plants soon saturates the air about them. Grassy surfaces become covered with dew.

The vegetation cover itself affects surface temperatures. In southern England, in summer, adjacent surfaces of sand and lawn gave temperatures of 55°C and 44°C respectively.

A plant's **response to night temperatures** may affect its distribution. Many plants can only flower, and therefore perpetuate themselves, when night temperatures are sufficiently low in relation to day temperatures. When night temperatures remain persistently high, they may not be able to reproduce successfully. The English daisy is an example of this: if the day temperature reaches 26°C, the night temperature should fall below 10°C. Quite small variations from the necessary night temperature may govern the variety and distribution of plants.

Fire

Fires, providing they are not too fierce, may clear obstructive litter and provide nutrients in the ash for a subsequent flush of growth of the dominant plants. Many temperate moorlands are seasonally fired to provide herbivores with nutrients from fresh, strong plants.

In some environments plants have become established only because natural fires have removed competitors which otherwise shaded out their seedlings. The giant sequoia in California finds its seedlings shaded out by other conifers unless the ground about the adult trees is regularly burned.

In Australia, in parts of the natural bush the cycling of material ceases if the organic litter is not fired at least once every fifty years. The grass tree *Xanthorrhoea* will only bloom after fire. The *Banksia* species are proteas which, like other Australian **pyrophytes**, only regenerate after fire, because otherwise their woody fruits will not open. Fires caused by lightning have thus played a major role in regulating the composition of Australian vegetation.

Light energy

Light energy required for the photosynthesis of carbohydrates is an important environmental factor. The intensity varies from place to place and seasonally according to the latitude. It is affected by the aspect of the plant habitat, by the proximity of other plant species, and by climatic factors such as cloud cover. Amid other leafy layers of

Fig. 9.1 The fire-blackened stem of a grass tree, *Xanthorrhoea*, in a semi-arid landscape in south-west Australia, standing amid the leaf litter of a lower shrubby vegetation.

vegetation, plants establish themselves at heights which will meet their light requirements, in relation to higher and lower foliage.

The time during which light and darkness succeed each other is often important. In **short-day** **plant species** flowering is promoted by long nights and short days. **Long-day plants** flower when the daylight hours are greater than a particular value. The leaves transmit messages to growing-points, and pigments in the leaves which absorb light of crucial wavelengths require periods of dark to regenerate to an effective chemical state. This, of course, means that particular plants will not be found where the daylight regime during the growing season fails to meet these requirements. It shows why scientists strive to produce crop strains which have an optimal leaf area relative to a certain length of daylight.

Air movements

These affect evaporation, transpiration, and the temperature of plants. Plant growth may be inhibited by too rapid wind-induced transpiration. The alignment of the branches on the lee side of a tree is often a response to the prevailing winds. And, of course, wind can structurally damage vegetation. Nevertheless, it is also an agent of dispersal, and assists the spread of particular plants.

Soil conditions

The balance between vegetation and soils is complex and often delicate. Deficiencies of minute amounts of a trace element in a soil may have a completely inhibiting effect on the growth of a particular species. Nutrient exchanges have been discussed (p.83), and in Chapter 10 we see the relationships between mature soils and the long-established vegetation of the major biomes.

Biotic factors

Soil micro-organisms have an immense influence

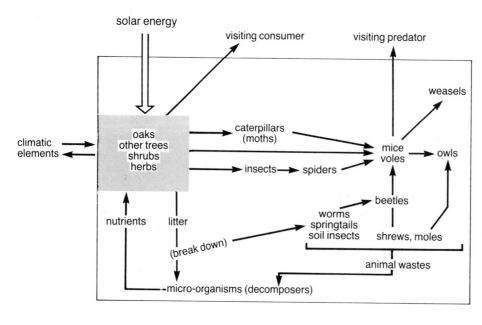

Fig. 9.2 The food web of an English oakwood, where energy passes through food chains, like that of oakleaves to caterpillars, to mice, to owls. The whole is much more complex, for the organic wastes are recycled by insects in the litter, which break down large material for the decomposers to convert to available nutrients, which are re-absorbed by the trees and herbs. Occasionally intruders, including people, or climatic changes affect the balance, which in time is readjusted.

on plants, and there are a multitude of other forms of animal life associated with any type of vegetation. As members of a balanced community in a biome, they exert controls and can bring about fundamental changes in the nature of the vegetation: surges of animal or insect population can cause dramatic changes in a relatively short period – witness the effects of locusts, caterpillars and rabbits.

Such sudden increases, and their consequent effects on plant life, are often related to a **changing balance between animal populations**. Birds of prey and large animals keep down the numbers of smaller birds and animals which, acting as seed carriers, are linked to the plant world. **Human interference** can also create chain reactions. These are not always as obvious as that following the mistaken policy adopted in China in the 1950s, when small birds were killed to protect grain crops: the huge increase in insect populations, formerly controlled by birds, soon reversed the policy.

Within any unit of vegetation there are numerous **food chains**, linked to form a **food web**. Fig. 9.2 looks at an English oakwood as a system, within which members of a temporarily balanced community form a web of food chains.

Changes in climate, or a population explosion of one of the life-forms, say the caterpillars, will upset the balance. In time this will adjust itself to a new, modified web, because some small creatures, deprived of food by the plague of caterpillars, may have died, and so have left the food web.

Some consumers, such as deer or cattle, and some predators, such as foxes, may temporarily enter the system; and, of course, people lurk threateningly in the background. Human interference with the vegetation is never confined to a single intended act. The destruction of particular trees, for example, interferes with all the activities of fungi, insects, animals, and other plant forms which are associated with them. They destroy a balance which can never be re-established.

9.3 Plant adaptations and competition

Dominant plants in the vegetation possess inherited characteristics which allow them to flourish in that particular environment. **In arid locations** plant organs are modified to obtain, store, and prevent the loss of water. Tissues are thick, surfaces waxy, leaves small or reduced to spikes, stomata contract and are protected. Deep tap-roots reach water far beneath the surface, elongating themselves as long as they remain in dry aerated conditions; 10 metres is not an unusual root length for a large plant. Succulents, such as cacti, store water in cells in the stem or leaves. Some plants spread roots over the surface to absorb as much moisture as possible during humid periods.

In a watery environment, as we have seen, many aquatic plants develop buoyant stems and leaves, and expose organs which allow air to pass through them, even the roots. Fig. 9.3 shows details of contrasting ways in which leaves are adapted to enable a plant to compete in a particular environment.

Inherited features must be efficient, or undergo further adaptation, if a plant is to survive the competition for space, light energy, and nutrients against the other numerous species in the vegetation. But, as indicated, *any form of vegetation must also be regarded as a plant–animal community*. Plants, micro-organisms and higher animals co-exist, maintaining complex inter-relationships.

Vegetation is made up of innumerable **micro-habitats**, and only close ecological studies can reveal whether or not a plant community is stable or undergoing modifications. At the macro-scale, few biomes can be regarded as having remained stable over the period of climatic fluctuations since the onset of the last ice age. Only the low latitude rainforests have survived the quaternary

Fig. 9.3 The pine leaf conserves moisture by its thick cuticle and fibrous layer beneath, by guard cells below the surface, and by resin ducts within the chlorophyll-containing cells. The water-lily is rooted, but the waxy leaf floats, with stomata on the upper side and slime glands beneath. The inter-cellular air space gives it bouyancy, and fibrous cells add rigidity.

PINE LEAF   xerophytic type

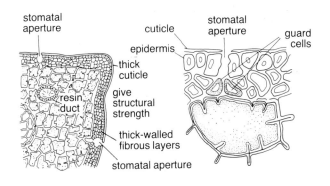

WATER-LILY LEAF   hygrophytic

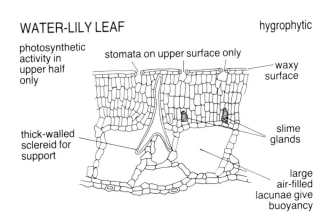

Fig. 9.4 Water-lilies rooted in a pond in Antigua. The floating leaves spread out and cover a wide area, their stomata exposed.

Fig. 9.5 The dry season on the East African savanna; the thorns of the young, leafless acacia stand out amid the seed heads of the stiff, dry grasses.

ice ages with relatively little change. As a result they have a high diversity of species. There are hundreds of tree species, and plant and animal species have their own particular stratified niches among the mass of competing life-forms.

9.4 Plant successions and climax vegetation

The composition of plant populations changes with time and generally becomes more complex. Lowly plants tend to give way to others of a higher form. Dominants at any particular time are replaced during a process of continuous change. This may continue until the plant community reaches **a stable state**, known as **the climax**. The composition of the vegetation may then remain more or less the same for a long period of time. However, this is usually only relative stability. Climatic changes, natural events – such as volcanic eruptions, drainage changes caused by tectonic uplift or depression and flooding – and, of course, human interference may once more initiate progressive or retrogressive change in the composition of the whole community.

Ecologists rightly regard the whole idea of plant successions leading to an ultimate climax vegetation as a convenience, for the sake of study, rather than an inevitable natural process. Nevertheless the concept of a climax community is an extremely useful one in the study of vegetation.

When plants colonise bare ground, the processes of dispersion and plant migration enable a number of families of particular species, or colonies of two or more species to establish themselves. They compete for available space, light, water and nutrients. As they live and die within

Fig. 9.6 When the rains come to the East African savanna, the leaves quickly sprout on the thorny stems of the acacia.

the area they help to change the existing habitat. They affect the micro-climate, the organic content of the soil, and its humidity and texture. As they struggle for existence they thus affect each other.

Other plants arrive to colonise the now-altered habitat, and in turn affect the existing balance: taller, sturdier species shade out lowlier ones, or deprive them of soil solutions. New dominants take over and exert their influence. Eventually a period of relative stability is achieved and a climax vegetation exists, maintaining itself in balance with the physical, climatic and biotic features of the environment; though there are bound to be minor adjustments from time to time.

Latitudinal influences exert strong controls over the distribution of biomes, through regional climates and soils. The highest type of vegetation which can exist in a region under particular climatic conditions is sometimes called the **regional climatic climax**.

Human interference in any natural region may impede the development of a regional climatic climax and permit only dominants of lower life-forms to establish themselves; as when forest clearance and seasonal burning maintains grassland communities at the expense of dominant trees. Such lower forms of vegetation may persist for long periods, for it may be impossible for the original vegetation to re-establish itself; the term **sub-climax vegetation** may then be used.

Some ecologists would regard any attempt at making a simple classification of vegetation on a world-wide basis as undesirable, or at least highly dangerous; partly because of the interference with natural successions, and partly because plants are only one of the life-forms present in a biome. Nevertheless, as we have seen, the major biomes are strongly influenced by temperature–humidity–soil relationships and acquire charac-teristics typical of a certain latitudinal spread. We consider the composition of these major biomes in Chapter 10.

First, it helps to be able to recognise the characteristics of certain units of vegetation, and to be able to describe the plant forms associated with them.

The sere: a series of plant successions

The development of vegetation on a particular site results in successive series of communities, each known as a **sere**, leading to a climax form.

Vegetation may have its origin in fresh water, salt water, damp surfaces, or dry ones. As higher life-forms develop, there is a tendency for the habitat to be modified by soil formation, humus accumulation, and moisture changes. This often results in hygrophytic plants giving way to mesophytic ones (p.103). In dry locations xerophytic plants may also be replaced by mesophytic ones.

A hydrosere. In the case of the aquatic environment, the development of vegetation may be initiated by bottom-rooted and floating water plants.

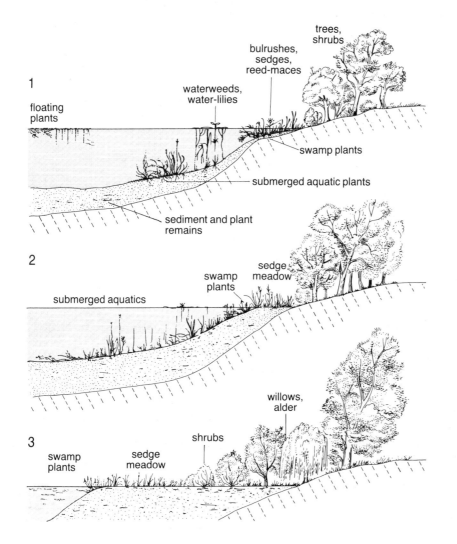

Fig. 9.7 Water plants and marsh plants gradually build up a bed of decaying organic matter in the silt. Consolidation and aerated conditions lead to the progressive establishment of marsh grasses, shrubs and trees.

They gradually build up a bed of decaying plant matter and silt, until, in shallowing water, swamp plants can enter the community and become dominant. Plant matter and other debris continue to accumulate, until a soil, with its humus content, is sufficiently consolidated and aerated to support coarse grasses, and, later, shrubs and trees.

Fig. 9.8 Salt-tolerant trees and shrubs with succulent leaves form a mangrove vegetation in a lagoon in coastal Antigua. Thin pneumatophores (breathing roots) grow up from the main horizontal roots; their minute openings allow air to enter, but not water.

Fig. 9.9 The remains of former mangrove vegetation growing in a silted-up lagoon. Low trees and scrub, zoned at first according to salt-tolerance and distance from water, tend to take over; low shrubs and grasses also establish themselves.

A xerosere. By contrast, a bare desert surface may first be colonised by simple life-forms which cling to, and help to corrode, rock surfaces. They add a small amount of organic matter to the rock debris, and this then supports incoming plants of a slightly higher life-form, capable of holding more water: lichens, for instance, may give way to small xerophytic herbs. Soil building and organic accumulation may continue, providing there is sufficient water in the system; so that hardy herbaceous plants may be followed by larger, woody ones, and perhaps, eventually, by a tree cover of sorts.

Such successions may be halted at any stage by unfavourable conditions. But they illustrate the tendency for higher life forms to become established, and for less extreme water conditions to result.

9.4 Plant classification

It is helpful to be able to describe plants according to their structure and the function of the various plant organs. In the following classification, for instance, the plants of the first three divisions do not possess the kind of structure that is found in plants of higher phyla, with cells forming tissue capable of distributing water, mineral salts and other nutrients via the roots, stems and leaf veins. Mosses, for example, attached by rhizoids and without true roots, depend primarily on photosynthesis for nutrition, but take in dilute solutions through their leaves.

Divisions (or phyla)

Schizophyta: minute organisms, like bacteria and simple algae.
Thallophyta: plants without properly differentiated roots, stems, or leaves; including large algae and seaweeds, fungi, moulds, and lichens.
Bryophyta: plants without true roots, but with the beginnings of stems and leaves.

Pteridophyta: plants with roots, stems, and leaves; with conducting tissues for the purposes described above, but reproducing through spores and not seeds. They include the horsetails and ferns.
Spermatophyta: seed plants with roots, stems, and leaves. In this category are most of the plants which make up the present biomes.

Classes: below the main phyla, plants are grouped into classes. Thus the phylum spermatophyta consists of **gymnosperms**, which are woody plants with ovules not enclosed in an ovary (conifers, for example), and **angiosperms**, which are flowering plants with enclosed ovules.

Orders; Families; and Genera: each class is made

up of various **orders**, usually with names ending in -**ales**. These, in turn, embrace **families** of plants with certain common characteristics, with name endings of -**aceae**. The families are grouped into **genera**, and the members of each **genus** have many obvious points of resemblance, and consist of one or more species.

Species: the plants in the same species are very similar to one another and are usually able to interbreed. The species are known by Latin names, so indicating the genus to which they belong together with a descriptive epithet. There are usually recognisable *sub-species* and *varieties* of plants within the species.

Not all authorities use precisely the same method of breaking down the classification; but the main pattern is followed, and is a useful one for accurate description. We can thus immediately identify a Scots pine on a hillside in northern Britain as a spermatophyte; on closer inspection as a gymnosperm, and as one of the most numerous orders of these, **coniferales**. We may next recognise it as a member of the genus *pinus*; and, if we have sufficient knowledge, identify it as a *pinus sylvestris*, or Scots pine. The final identification of a plant in this way may call for considerable experience.

Raunkiaer's Classification

Another type of classification, first made by the Danish professor C. Raunkiaer, enables us to describe the composition of the vegetation in terms of broad groups of life-forms.

(a) Perennial shrubs and trees with stems and renewal buds more than 250 mm above the soil and exposed to most climatic hazards (**phanerophytes**).
(b) Perennial herbs and low shrubs with renewal buds between ground level and 250 mm above: plants like thyme and the saxifrages (**chamaephytes**).
(c) Herbs and grasses with resting buds at ground level or in surface soil: plants like dandelions and nettles (**hemicryptophytes**).
(d) Plants with underground bulbs, tubers, or rhizomes well-buried in the soil: such as crocuses and hyacinths (**geophytes**).
(e) Plants with a full life-cycle under favourable conditions, but which survive later stringent conditions, say of aridity, in the form of resistant seeds or spores (**therophytes**).
(f) Water plants and marsh plants (**hydrophytes** and **hygrophytes**).
(g) Plants which grow in debris on host trees, using them as a means of elevation to the light they require for photosynthesis, but without harming the host (**epiphytes**).

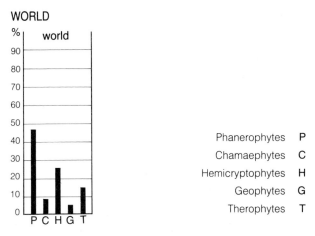

WORLD

Phanerophytes P
Chamaephytes C
Hemicryptophytes H
Geophytes G
Therophytes T

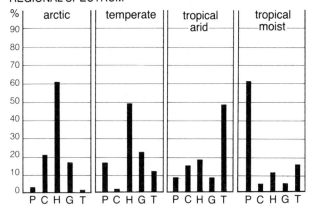

REGIONAL SPECTRUM

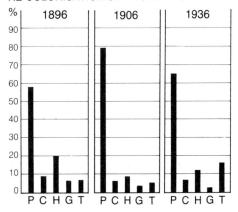

RE-COLONISATION OF A VOLCANIC AREA

Fig. 9.10 **(Above):** the composition of the vegetation in very broad zones. **(Below):** the spectra show the colonisation of a relatively small area in Indonesia where volcanic activity had destroyed vegetation; revealing, among other things, the effects of competition among the various plants.

Fig. 9.10 shows how this simple system can be used to present a broad biological spectrum of the percentage of life-forms described to the total flora in any region, large or small. It can also be used to show the changing composition of the flora with time, such as the recolonisation of a disturbed area.

Grouping on the basis of temperature requirements

Plants may be classified in terms of the mean temperature limits within which they can flourish. The following classification considers mean monthly temperatures.

(a) All months above 18°C (**megatherms**).
(b) Coldest months 6–18°C; warmest months above 22°C (**mesotherms**).
(c) Coldest months above 6°C; warmest months 10–22°C (**microtherms**).
(d) Warmest months below 10°C (**hekistotherms**).

Such groups are extremely broad, and as plant environments vary from one tiny locality to the next, can have only limited use in description.

Vegetation units: classified by status

Large units of vegetation include a variety of different plant communities; so it is often necessary to refer to smaller groups of plants within the larger unit. In terms of status, these can be classified as follows.

Formations: these are large distinctive units such as tundra, coniferous forest, or heath.
Associations: these are various climax vegetation units, each with its own characteristic dominant species, which together make up the formation. Thus a mixed deciduous forest formation may include an oak-beech association; or one with a single dominant, as in a beech wood on chalky soil.

Fig. 9.11 Thallophytes – lichens and mosses – take hold on Vesuvius lava. Plates 14 and 15 show progressive stages of colonisation.

Societies: these are groups of lower life-forms in which one or two species predominate because some particular feature of the environment especially suits them – a certain type of soil perhaps. Plants of several colonising species may be grouped together within a larger vegetation unit. On a smaller scale, groups, or clans, of a single species – or even a small group which has spread from a single invading plant – may have local dominance.

Fig. 9.12 Cacti and aloes growing in volcanic ash on Lanzarote. They are flowering xerophytes with roots, stems and leaves. But in the case of the cacti, the leaves are reduced to tiny, hard-pointed prickles on a flat, water-storing stem which performs the functions of respiration and transpiration. Here their growth is encouraged to feed the cochineal beetle, from which dye is obtained.

10

MAJOR BIOMES AS ECOSYSTEMS

10.1 The biomes

Climatic influences are the chief controls in the distribution of biomes. As climate and vegetation together strongly influence soil formation, so the large-scale soil distributions correspond closely with the major biomes. Biomes are recognised mainly by the dominant life-forms of the mature vegetation: but they are, in fact, **ecosystems** in which these dominants influence the habitats of other plants and of the numerous forms of animal life with which they are associated.

The distribution maps of these biomes are, of course, largely unreal, in view of human activities. Most show what the extent might have been before people interfered. They refer to more or less mature vegetation formations, most of which are the culmination of vegetation successions related to post-glacial climatic adjustments. The main areas to have survived with relatively little alteration are occupied by tundra and low latitude rainforests; and the latter are shrinking at an alarming rate.

Fig. 10.1 The low latitude seasonal forests vary considerably in their plant associations, structure, and behaviour.

10.2 The low latitude rainforests

As Figs 10.1 and 10.2 indicate, these mainly cover tropical lowlands, where all-year-round precipitation and high temperatures cause a great increase in the weight of organic matter during the year. **The rate of accumulation of biomass** in unit time is called the **Net Primary Productivity** (NPP). In the case of these rainforests this averages some

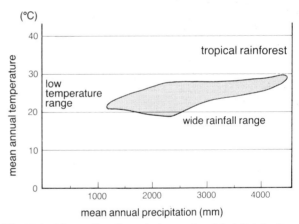

Fig. 10.2 Macro-climatic controls over the global distribution of the biome.

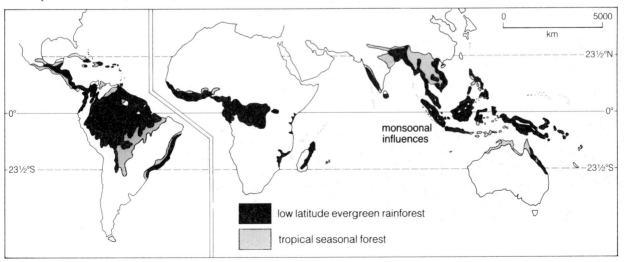

low latitude evergreen rainforest

tropical seasonal forest

23 tonnes of dry matter per hectare each year (compared with 13 tonnes in temperate deciduous forests and 8 tonnes per hectare in the northern coniferous forests).

Diversity and interdependence within the ecosystem

These are dense forests with **a high diversity of species** and luxuriant growth. Fig. 10.3 shows that there are no soil-water deficiencies. The forest is evergreen, the species deciduous. That is, despite the lack of seasonal temperature variation, individual plants lose their leaves at intervals, even though there is no general period of leaf-fall, as there is in the temperate deciduous forests. The trees of certain species lose their leaves at about the same time. But among the plants as a whole, tissue growth, flowering, fruiting and leaf-fall go on all the time, so that the overall impression is of a close green forest.

The impression of a complete evergreen cover is

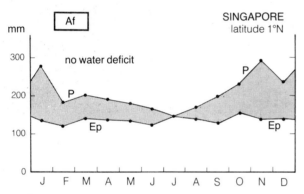

Fig. 10.3 The monthly precipitation always exceeds evaporation losses.

reinforced by aerial views. But, in fact, **the vegetation is stratified**. Huge trees, 40–50 metres tall, often widely spaced, rise irregularly and form the upper emergent zone. They project through a lower, much more continuous canopy of trees, whose crowns form a layer some 25–35 metres up. This layer absorbs some 70–80 per cent of the incoming **light energy**.

Beneath are shorter, more slender trees with narrower crowns, their heights varying from about 5–15 metres. The terms 'strata' and 'layers' are misleading, however, for amid the vegetation there are always young trees of the higher species, and a mass of branches and foliage, with hanging lianas and creepers festooned about the trunks. Only at the lowest level, where little light penetrates, is the forest relatively open and free from higher forms of ground vegetation. At the edge of rivers, in clearings, or on hillsides, where light penetrates more readily, there is often a dense lower growth, with ferns, grasses and shrubs. Even though the forest floor receives only about 1 per cent of the energy arriving at the crown, the flecks and flashes of more intense light provide considerably more energy than the actual measured shade light. This helps the development of young plants, saplings, and large herbs, such as the various banana species.

The tops of the emergent trees receive most of the energy. Many of the abundant leaves of the crown swivel in relation to the sun's position. The amount of energy absorbed depends on the leaf's albedo, and some comes from reflection from other leaves. Energy is lost by radiation and convection, and as latent heat through evapotranspiration, depending on the temperature and air movements. As the leaves are alternately exposed to rain and drying conditions, most have a thick

RAINFOREST STRIP
8 metres wide × 60 metres long

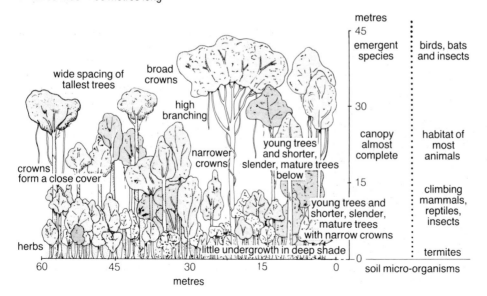

Fig. 10.4 The profile gives an impression of ragged vegetation with huge mature emergents, and younger trees and shorter species beneath. But from above, the surface seems to be completely masked by the vegetation. The animals, too, occupy their own particular vertical zones and niches within the forest, and species are also separated by preferential daytime or night-time activities.

cuticle; in general, the hotter and wetter the conditions, the larger the leaves. The life-span of the leaves varies from 4 to 14 months. Many have long, pointed, channelled tips which shed water, for when transpiration is checked so is the process by which the soil water and nutrients are drawn up. Many of the large-leaved herbs of the lower levels exude liquid water by guttation, allowing internal movements to continue.

Epiphytes abound in the upper parts of the tree canopy, and lianas, with their pliable, woody stems attach themselves to the trees as they wind upwards to the light. Different species of epiphytes and numerous parasitic plants also flourish at lower levels. Epiphytes, such as the bromeliads, equipped with water-absorbing scales, can only take in water during rainy periods; many of them therefore have succulent water-storing leaves, as their roots are only a means of attachment. Epiphytic humus, however, provides nutrients for other plants, such as the ferns which are rooted in litter in the forks and cavities of trees. Dripping water and dust bring in other nutrients. Seeds in ants' nests can develop into flowering plants, a reminder that the forest biome is a balanced community of animal and plant life.

The forests are never without flowers, though many cannot be seen from the ground. The intervals between **flowering** are as varied as the flushes of growth and the time of leaf-fall. There is rarely pollination by wind; for one thing, with such diversity of plants the individual species are widely spaced throughout the rainforests. Insects, bats, birds, and mammals all play their part in pollination. The interdependence of organisms, and the delicate balances within these communities, are shown by the huge silk–cotton tree, whose flowers open at dusk, when flocks of bats come to lick the nectar and carry away pollen on their fur. Each flower loses its petals before the next night.

Fig. 10.5 Drip-tip leaves rapidly shed surface water.

Fig. 10.6 A tall emergent festooned with vines, lianas, and epiphytes rises from the middle canopy of a Caribbean rainforest.

The plants are the primary producers of organic matter, consumed by innumerable other organisms, from caterpillars which strip the leafy crowns to termites which break down woody tissues on and beneath the forest floor – where mites and insects reduce the litter and where fungi cause rapid decay.

The larger animals occupy ecological niches which suit them best. More than half the mammals are at crown level, many with prehensile tails. There is stratification within the forests, but it is blurred by the mobility of the insects, reptiles, birds, and mammals. Birds and bats find insects and fruit in the crown, and squirrels and monkeys consume plants and insects at various levels. Some are adapted to glide between trees and so increase their range. The climbing cats have a mixed diet, and like other carnivores or omnivores can descend to ground level where other mammals, such as pigs, find food.

Nutrient-cycling: the delicate balance

Such dense interdependent communities give the impression of a biome of immense fertility, located

Fig. 10.7 The flanged trunk of a tall silk-cotton tree in a Central American rainforest, showing the flecks of light which are so important to vegetation in the lower levels of the forest.

Fig. 10.8 Various epiphytic plants compete for positions which will give them a suitably sunlit site on the tree.

where the heat and warmth are bound to foster abundant life-forms. In fact these productive ecosystems flourish on soils which are mostly very old and deeply weathered, but poor in nutrients, with pH values of the order of 4.5–5.5.

On the forest floor, where the spreading root buttresses support the very large trees, litter arrives at the rate of more than ten tonnes per hectare per year. The whole complex system depends on **the recycling of the nutrients** from this rotting biomass. Intense biological activity is involved. Shallow, spreading plant roots penetrate the litter layer, and are associated with the fungus *mycorrhiza* which, on breaking down organic material for its food, releases nutrients which are absorbed through the plant roots. The litter decays very rapidly, and colonies of termites help to break down woody tissues. There are also nitrogen-fixing micro-organisms; nitrogen is introduced into the system by rain, and the nitrogen cycle is particularly important in sustaining the biomass. Other nutrients such as potassium, silicon and calcium are also important and are continually recycled. As a result **few nutrients are lost through leaching**, and, in fact, the forest streams are usually very low in dissolved minerals.

The dangers of *interfering* with these processes are obvious. The infertile ferrallitic soils will not in themselves provide nutrients for forest regeneration when cleared of vegetation. They physically deteriorate when exposed to tropical downpours, and the formation of impervious duricrusts becomes likely. Clearance for shifting cultivation by a limited number of people, who abandon the plots after a few years to allow vegetation to recolonise, allows nutrient cycles to recommence; but the balance of the natural species is never the

same, and huge areas of the present rainforests are, in fact, secondary vegetation, especially in south-east Asia. The danger with any clearance is the loss of not only a particular plant species, but the disturbance of all the other life-forms associated with it, p.154.

Fig. 10.9 A rainforest biome seen as a simplified system, with exchanges of energy and matter which make for finely balanced nutrient flows.

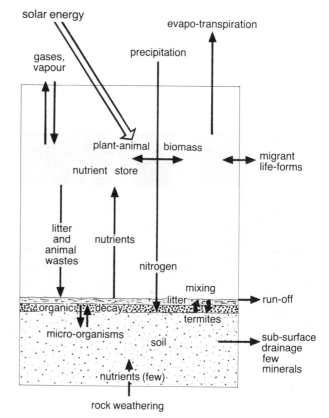

10.3   The tropical seasonal forests

Where there is a pronounced dry season, the tropical forests have different species and different characteristics. They occur particularly in the monsoon countries of southern Asia, West Africa, and poleward of the equatorial lowlands in South and Central America.

In these areas annual rainfall may be smaller, but after a dry period of several months there is a period of heavy, prolonged rain; and in some monsoon locations the annual total is among the highest in the world. There are greater seasonal changes in the length of day and light intensity than in the low latitudes, and annual and diurnal temperature ranges are usually higher.

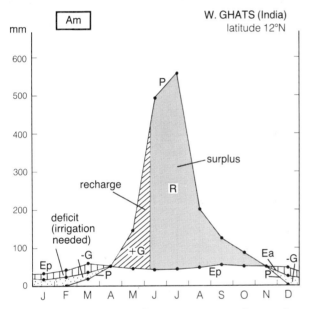

Fig. 10.10   During summer the precipitation greatly exceeds the actual evapo-transpiration. Once the shortage is made up, a soil-water surplus is created. This is lowered during the dry season. The natural vegetation responds by deciduous behaviour and other means of conserving moisture, but loses an amount, Ea, by evapo-transpiration. Ep − Ea represents the deficiency (D) that would have to be made up by irrigation to allow maximum growth.

Plants must survive a period of soil water deficit, which varies in length according to location. The vegetation varies from luxuriant forest in well-watered regions with a relatively short dry season, to drier deciduous forest, and even open grassy woodland. The wetter forest contains many deciduous emergents, tall trees which shed leaves in the dry season, though lower levels have a high proportion of evergreens. In the drier forests deciduous trees are the main species, with a lower discontinuous layer of evergreens. In the dry parts of northern Burma and eastern Assam tall hardwood such as teak, sal, and in are deciduous dominants, and have become commercially important; they show seasonal growth-rings in their wood. They are still massive, but do not have the buttressed base of many rainforest trees. There is more branching, and more trees with spreading crowns. There are fewer climbers and epiphytes. Flowering is most common during the dry season.

Where there is seasonal leaf-loss, where light slants onto hillsides, or wherever an incomplete leaf canopy allows light to reach the ground, the undergrowth is very dense. In places thick shrubs and clumps of tall bamboo form almost impenetrable thickets.

The decomposition rate of litter is usually slower than in the low latitude rainforests, and the clay−humus content of the soils more stable. Dry litter can add to the intensity of forest fires and so exclude some species found in evergreen forests: an important ecological factor.

10.4   The vegetation of tropical shores and coastlands

Where sediments accumulate near the shores, as deltaic deposits or mudflats, and where there is tidal salt water, or brackish water, mangrove forests and swamps often form a thick fringe of green or greyish vegetation.

Various species of **mangrove** trees form the main dominants of this **halophytic evergreen vegetation**. In many of these trees suction forces produced in the leaves are transmitted to the

Fig. 10.11   Mangroves with stilt roots line an East African coastal creek. They drop seedlings which penetrate the mud and produce lateral roots in a matter of hours; some seedlings are carried by the tide to root on sandbanks.

roots. When this is greater than the osmotic pressure of saline solution it prevents excessive salt entering the plant and building up. Other species have different protective mechanisms.

Together, the trees of low to moderate height form an apparently tangled mass. There are few other vascular plants, drawing solutions through their systems.

Many of the mangrove trees have sharply arched prop-roots projecting from the mud. Others have spreading roots, and branching from these are narrow, conical projections capable of taking in air – the pneumatophores seen in Fig. 9.8. Stems and roots bear reddish algae and various lichens.

The tangle of roots helps to build up silt, so that, as the surface consolidates, other forms of mangrove may develop. Sometimes they are **a stage of a hydrosere** which develops from hydrophytic to mesophytic forms of vegetation (p.107).

Thick, leathery, shiny leaves guard against excessive transpiration loss. In many species seeds germinate in the fruit while still on the plant. As the seedling breaks through, it extends downwards like a dagger, eventually falling and embedding the radicle in the mud, to form anchoring roots. Others are carried away by water, perhaps to root elsewhere.

On firmer sands the roots of coarse grasses and other halophytic plants help to bind the surface. The pandanus, or screw-pine, with its clusters of strut-like roots, and tufts of thin dry leafy blades on a woody stem, is a typical tree-form along tropical coasts. While on firmer shores low, woody evergreens with dark green, glossy leaves form a lower layer beneath shallow-rooted coconut palms. The coconut fruit, with its thick fibrous layer and hard shell, can withstand a fall of over 20 m, and, being buoyant, can be dispersed by floating to distant shores.

The animals are a mixture of land- and sea-creatures, exemplified by the mudskippers, which travel over exposed mud on modified fins. Fish abound in creeks. Barnacles encrust roots; crabs adapt themselves to zones of vegetation; insects breed in stagnant water; and birds and monkeys move through the branches; the larger predators include cats.

10.5 Tropical grassland-with-trees (savanna)

The distribution of this form of vegetation, shown in Fig. 10.12, appears to fit neatly into the zonal pattern based, primarily, on climatic variations: a zone where, by definition, grasses are dominant, but associated with relatively abundant trees in the wetter areas, and with thorn-bush scrub nearer the drier margins.

This is far from reality. In the llanos of Venezuela huge tracts of grassland, with small, scattered trees or patches of low woodland, cover hot, wet lowlands, which receive more than 1300 mm of rain a year. There tropical seasonal forest would seem the more likely form of vegetation. In northern Tanzania patches of dry scrub grassland lie within tall grass parkland, whose scattered trees are ten metres or so tall.

A variety of plant associations is found throughout the tropical grasslands-with-trees; though overall there is a trend from woodland to semi-arid scrub with increasing latitude, and towards the interior of the landmasses shown in Fig. 10.12. The variations appear to be responses to a number of quite separate controls:

(a) climatic; particularly the length of the dry season.
(b) soil conditions (edaphic).
(c) fire caused by natural events (lightning) or by people (pyrogenic).
(d) grazing by wild herbivores or domestic herds.

Fig. 10.12 This broad distribution includes a great variety of grass-dominant associations.

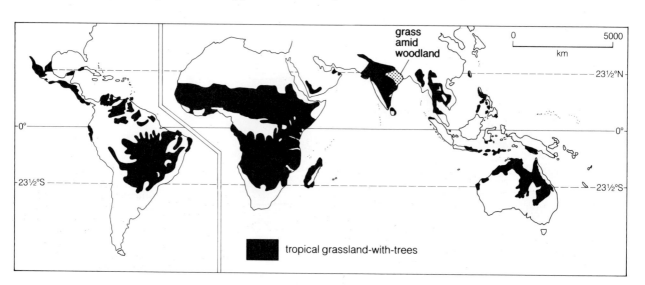

Climatic controls involve the availability of soil water through the year (Fig. 10.13). Grasses and woody species tend to exclude one another.

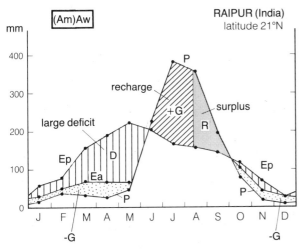

Fig. 10.13 A station with a single rainfall maximum. During the long dry season there is less than 30 mm of rain a month, so by May there is a large water deficiency. The short, hot wet season creates a surplus; but high temperatures continue and this is soon withdrawn. Plants must adapt to very low soil-water levels.

Grasses have a close, finely-branched root system. They flourish particularly in fine, light soils with the ability to hold water during the dry season. Woody species have coarse roots capable of extending into the soil in any direction, penetrating stony soils and drawing water from considerable depths, though sometimes hindered by plinthite layers.

With the rains the grasses grow very quickly and transpire freely. Transpiration continues until the aerial part of the plant dies, leaving only the root system and the terminal growing point. Growth begins with the first rain.

Woody plants respond to drought by closing stomata and reducing transpiration. Trees have different responses to a lengthy dry season. Some retain small, hard xeromorphic leaves, others have larger leaves which they shed during the dry season and develop again with the rains. Though many species shed leaves, the woody stems and branches continue to lose some water and require a small supply. Thus, in really dry areas, where grasses use up the water reserves of the top-soil it is difficult for young woody plants to survive.

Where rainfall and soil water storage is sufficient for trees, a shade factor becomes important. If the tree crowns form a canopy, the grasses must adapt to the poor light conditions at ground level, and the woody plants may be the dominant competitors.

Edaphic controls often relate to the amount of water available from the soil. But there are many reasons why a great variety of soil conditions may occur over a single savanna landscape. Many of the savannas cover very old land surfaces; so that, in broad terms, the grasses and trees grow in ferrallitic soils, most of them depleted of mineral nutrients. Some are on soils underlain by plinthite concretions at various depths (Fig. 8.15), or in some cases grow on a duricrust surface; others have to cope with the seasonal contractions/expansions of vertisols (p.99).

On a local scale, the continuous development of

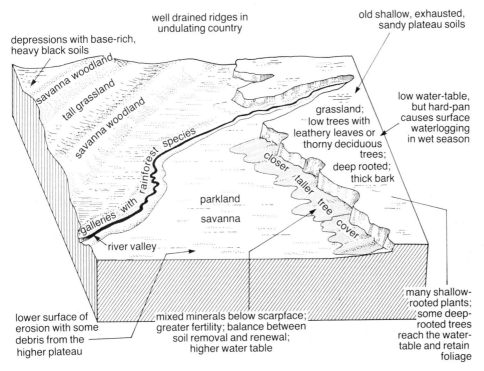

Fig. 10.14 A dissected tableland under *Aw* climatic conditions, with hot season rainfall of some 700–1400 mm. Geomorphic features and variations in the depth of the water-table make for a variety of soils and vegetation in an area of some 100 square kilometres.

Fig. 10.15 Maribou storks rest on an acacia in full leaf. This is the hot season in southern Kenya, when water drains into this lowland. With a high water-table, there is abundant grazing. However, blackened tree trunks suggest that fire swept through this area during a recent dry season.

landforms produces a variety of soils, like those related to the catena in Fig. 7.11. These relatively infertile soils on old duricrust surfaces give way to those below scarp faces, which are enriched by mineral mixing and lateral drainage. Fig. 10.14 shows how evolving landforms may affect the composition of the grassland-with-trees vegetation.

On a still smaller scale, termites are again responsible for mixing the upper soil; and **termitariums** are of such proportions and frequency in some savannas that the landscape is dotted with trees rising from island-like termite mounds (Fig. 10.18).

Fires sweep rapidly across the savannas during the dry season. Lightning, in particular, causes many natural fires; while on the African savannas, especially, large areas are fired seasonally to encourage the growth of grass species with a high nutritive value for grazing animals.

Bush fires consume litter that otherwise would have enriched the soil with humus, and tend to dry out the upper soil. On the other hand, minerals from the grasses and shrubs which are burnt, but mostly survive the fire, enrich the soil as nutrients. As Fig. 10.16 shows there is continuous recycling of nutrients between grasses and soils. Burning releases potash for the soil, but in a readily soluble form. Rain, of course, may eventually wash away many of the nutrients from the ash; and carbon, hydrogen and nitrogen compounds are lost in gases and smoke.

Most savanna trees with thick bark are fire-resistant once they are established, as are many herbaceous species. In Australia hundreds of species of evergreen eucalypts are adapted to the varying water resources of savannas which range from extremely dry to seasonally very wet locations; adaptations which allow them to survive

include oils in stems and leaves, twisting leaf stems which keep the leaf-edge to the sun, and deep roots. Most have fire-resistant seed cases. The dry vegetation associations vary in different parts of Australia; many are dominated by species of acacia. We have seen that in some Australian regions, the cycling of material comes to a halt if organic plant-remains are not burnt off after a period of years: this affects the *Banksia* and the spiky grass-trees, *Xanthorrhoea* (Fig. 9.1).

Fig. 10.16 The savanna grasses need a high uptake of silica, and continuous cycling takes place between silicate rock minerals and the vegetation. Biochemical actions, initiated by root exudates, help by making the element available in soluble form. The annual processes of decay leave silica in very small particles, which are much more soluble than that from rock sources.

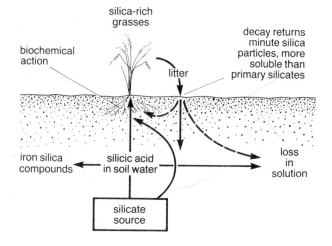

Fig. 10.17 A wildebeeste herd migrates across open grassland in southern Kenya, which is kept free of trees and shrubs by grazing. The hill behind, not visited by the vast numbers of ungulates which cross the plains, bears a close cover of trees and shrubs.

The animal life of this complex biome is affected by fire in different ways. Reptiles bury themselves, mammals and birds flee the flames, and the latter often thrive on winged insects doing the same, though many invertebrates perish.

The animals themselves use different levels of vegetation: giraffe take the top leaves of the acacia, rhino the lower shoots; while antelope, wildebeeste, zebra and many others graze rather than browse. There is a wide variety of scavengers and predators, with interlocking food-chains. The same plants provide food for many different species, though some animals are very selective.

As the position of the ITCZ changes through the year, there are mass migratory movements of herbivores and predators, broadly following the rains and the grass growth. Animals help to favour grass species by dispersing seeds, passing them through their system and carrying them on their coat. In general, their grazing prevents the development of tree shoots, but their consumption of fallen acacia fruit helps the plant to spread.

Unusually dry periods can affect grasses over a wide area and lead to overgrazing. This can be particularly harmful when the drying of water sources concentrates animals in the vicinity of certain water holes. The dearth of selective food-plants may affect particular species of animal. Herdsmen normally keep their cattle on the move; even so, the grassland can become overgrazed about the permanent settlements. This is made worse when normally migrant herdsmen move in during a drought or when their usual migratory routes are restricted.

Fig. 10.18 The East African savanna during the dry season with white, parched grasses and leafless trees. Here snakes live in holes in an old termite mound. The weaver birds build their nests as far away as possible from investigating snakes. The ground around the mound itself has been pounded bare by the feet of mongooses looking for snakes.

Fig. 10.19   A top-feeder among whistling-thorn savanna.

Fig. 10.20   A consumer digests its kill – a giraffe.

Visitations from locust swarms, which occur as periodic outbreaks from dry breeding areas, affect both the natural vegetation and the human occupants of the biome.

It is instructive to compare the primary and secondary productivity of the tropical grasslands with those of the tropical rainforests. The mean quantity of organic matter created through photosynthesis (**the primary productivity**) is some 2200 grammes per square metre a year in the rainforest, but only about 900 grammes in the savannas. **The actual plant biomass** in the forest is of the order of 450 tonnes per hectare; in the savanna it is some 40 tonnes, and most of the biomass is underground.

However, considering the low rainfall and poor soil qualities, the savannas have a high biological productivity. There are huge populations of large animal life. Herbivores make up about 95 per cent of the animal biomass.

The **ecological diversity** of the ungulates, the gazelles, wildebeeste, zebra, etc., enables them to make optimum use of plant growth. Each, as we have seen, has its own ecological niche, with a favoured diet: giraffe take top foliage, rhinoceros from bushes, and wildebeeste grass.

So the **secondary productivity** of animal matter in the savannas is estimated at some 200 kilogrammes per hectare per year, compared with about 150 kilogrammes in the rainforests.

In general the food chains in the tropical grasslands are simple ones, the most common being grasses → ungulates → carnivores. Fig. 10.21 shows how scavengers are also involved.

Sometimes dietary quirks lead to unusual behaviour: in Fig. 10.22, for example, elephant overpopulation has led to a shortage of a trace element which the elephants can obtain by eating bark; and in so doing they cause considerable damage to the trees.

Fig. 10.21   Other consumers move in on the abandoned kill.

Fig. 10.22   Elephants do a great deal of damage in the tree savanna. Their growing population is of concern in some protected reserves. During the dry season they eat tree bark to obtain the trace elements they need, and uproot trees or tear away branches.

10.6   Biomes of the sub-tropical arid and semi-arid regions

In each of these dry regions **there is a soil water deficit for most of the year**, as in Fig. 10.23, for a Saharan location. The storms of the short wetter period occur when temperatures and evaporation rates are high. The amount of water retained is insufficient to create a surplus. Yet some form of vegetation is found, however sparse, in most of the world's arid lands: so water must, at least, periodically, be available to plants. In the drier areas plants and animal life are confined to small, scattered, favourable micro-habitats.

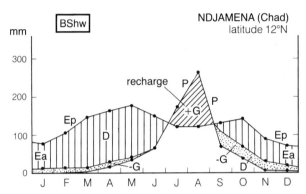

Fig. 10.23   This Saharan location has a large soil-water deficit for all but the late summer months. Although summer rains may recharge the soil-water content, there is never a surplus. What this does not show is the erratic nature of the rainfall, nor that several drought years may occur in succession (Fig. 11.12).

Much depends on the nature of the rocks, regolith and soil, skeletal though they may be. In humid regions sands bear relatively sparse forms of vegetation. However sands allow water to soak through them, so that the lower parts remain moist for a long time. Thus in hot dry locations they bear more vegetation than adjacent clays, which keep water near the surface, from where it dries out. A cracked, rocky surface may also allow water to penetrate deeply enough to be retained for long periods, and to be made available to plants with long penetrating roots.

In these regions **a large proportion of the biomass is beneath the ground**, for many plants have branching root systems, so as to draw water from the greatest possible volume of soil. Others have roots which branch laterally over a very wide area, but no more than 30 mm below ground, and so exploit every shower to the maximum.

Having acquired water, the plants are adapted to reduce its loss. Many have small leaves, which they can shed. Those with succulent leaves and stems have sunken stomata, and close their openings during the day to reduce transpiration. They open them at night to allow respiration, and to take in carbon dioxide for the subsequent photosynthetic production of organic matter during the day. Most of these succulents have small absorbing roots which die down during drought.

Some plants have pale leaves with reflective surfaces. Thick leaf cells with waxy deposits help reduce water loss. In cacti branches are reduced to small protuberances and leaves to spines. In other plants the stomata are mainly on the underside of the leaf. Some desert grasses have folded leaves with the stomata inside; while others have leaves rolled into tubes.

Fig. 10.24   Root systems adapted to occasional rain and a low water-table: the Rose of Jericho by its 'tumble-weed' behaviour; the grass by minimising loss through transpiration.

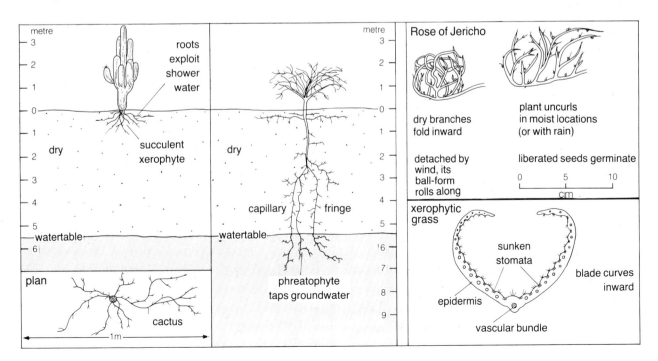

ARID/SEMI ARID REGIONS - VARIATIONS IN VEGETATION

B————B   boundary of arid/semi-arid 'B' climate

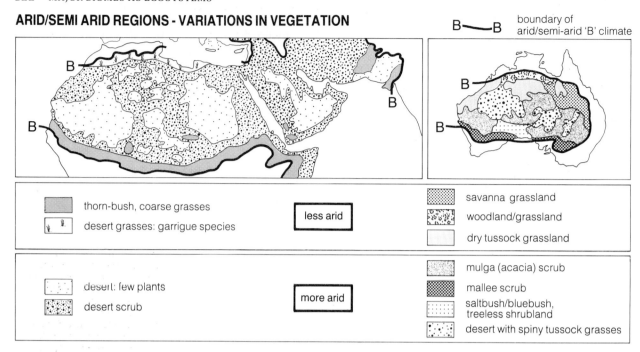

▨ thorn-bush, coarse grasses	
↓ desert grasses: garrigue species	**less arid**

savanna grassland	
woodland/grassland	
dry tussock grassland	

⬚ desert: few plants	
⬚ desert scrub	**more arid**

mulga (acacia) scrub	
mallee scrub	
saltbush/bluebush, treeless shrubland	
desert with spiny tussock grasses	

Fig. 10.25   For semi-arid lands, as with the savannas, variations from one small locality to another means that a generalised map has serious limitations. Vegetation with plant species adapted to arid conditions can be found in regions designated 'savanna'; while in the heart of a desert there are often surprisingly large areas which support ungulate grazing.

Certain plants are ephemeral, completing their life cycle during a rainy period, and then remaining dormant in the soil as fruit or seed. Those in favourable localities scatter seeds about them; others are dispersed by the wind. In some cases the whole dry plant-mass breaks loose and tumbles across the landscape, perhaps to rest in a moister hollow favourable for seed germination. In semi-arid areas goats and other animals help dispersion by carrying barbed seeds on their coats.

Seen on the broader scale of **vegetation distribution**, it is tempting to assert that as the rainfall amounts decrease, and the period of summer rain becomes shorter, tree savanna gives way to ever more widely spaced tussock grasses, low thorny trees and xeromorphic shrubs. Certainly thorn bush and scrub characterise such semi-arid areas as the dry north-eastern shoulder of Brazil. But, as with the savannas, there are abrupt contrasts in the degree of aridity and the vegetation responses in what are classified overall as semi-arid climatic conditions. Along water courses, dry for much of the year, and in shallow, almost imperceptible, drainage basins there are often tall grasses, numerous acacias, and isolated baobabs, storing water in the tissues of their bulbous trunks; yet the gentle interfluves may bear only dry tussock grass, with succulents such as spurges and cacti.

These forms of vegetation are easily disturbed, and may be affected by people's activities. Here some of the most drastic results of overgrazing have combined with the natural hazards of periodic droughts of unusual severity to encourage **desertification**; that is a deterioration of soil conditions, accompanied by the replacement of dry savanna and semi-arid plant forms by xerophytes and halophytes (salt tolerant species).

Fig. 10.26   Cacti on a sandy spit in Kingston harbour, Jamaica. The closely wooded hills beyond show that climate alone is not responsible for soil-water deficiencies and plant adaptations of this kind.

Plate 7 Waterlogging amid reclaimed farmland in arid country bordering the lower Indus. The surface water evaporates rapidly, but is continuously fed by groundwater from irrigated areas up-stream. The water becomes progressively more saline.

Many extensively damaged parts of the earth's surface, and those with increasing desertification, are the result of a lack of understanding of ecological balance by local people unaware of the causes. But even those who undertake sophisticated engineering projects may be ignoring factors which will ultimately feed back and work against their original intentions.

In Plates 7 and 8 we see how, despite very successful irrigation projects, involving long canals fed by the Indus and its tributaries, areas of potentially fertile land became a salt-caked waste.

Here perennial irrigation transformed the desert to farmland. But excessive watering upstream and percolation from the 'tails' of canals caused waterlogging. Groundwater moves downstream above clayey horizons and comes to the surface, allowing salt to be concentrated by evaporation.

Plate 8 Evaporation has left salt on the surface of large areas formerly converted to cultivation. Halophytic shrubs have established themselves on what was fertile farmland.

Plate 9 Inversion over Rio de Janeiro traps a misty layer, brown with urban pollutants and fumes from the numerous vehicles.

Plate 10 In East Africa many Maasai are now living in semi-permanent settlements; the state of the soil and vegetation reflects their impact on the local environment.

Population concentrations and environmental impacts vary with size and technology. In Plate 9 the activities of millions create a polluted atmosphere; in Plate 10 relatively few Maasai have stripped trees and bushes for fuel, leaving an exposed surface.

Fig. 10.27 Salt-bush and blue-bush of the Australian semi-desert are nutritious plants in a dry landscape which supports surprisingly large numbers of kangaroo, emu, rodents, snakes, lizards, flies and other insects. They provide grazing for extensive sheep farming. The individual shrubs need adequate ground-water, and are well spaced, with bare earth between, as in this salt-bush country, about a prospect mine in Western Australia.

Fig. 10.25 shows how, on a continental scale, controls due to the climate, soils, and drainage conditions produce mosaics of different forms of vegetation. Australia is without the thorn bush, but in the arid areas dwarf eucalypt species, known as mallee, give the appearance of twisted thickets, as several stems rise from each underground stock; and some species store water in the roots. In other parts of the interior low acacia trees and bushes are part of a plant association known as mulga scrub. This includes salt-bush and blue-bush, which with the edible acacia can support sheep in surprisingly dry country.

Plant and animal life respond rapidly to rain. In the hot deserts showers of rain cause seeds which have lain dormant, perhaps for decades, to germinate and develop. The annuals rapidly bloom during their short life-cycle, and hidden geophytes spring up and flower. Otherwise, it is only around wells in oases, or on irrigated flood-plains of rivers which rise beyond the boundaries of the arid lands that there is abundant vegetation, with date palms, food crops and fodder plants.

As rain triggers an unusual flush of vegetation, so an abundance of animal life appears. Insects visit the flowers; caterpillars and crickets eat the ephemeral plants; flies provide a rich food supply for birds, which arrive in large numbers. Spiders, scorpions and other surface predators feed on the numerous small creatures busy with their own responses, such as the ants harvesting grass seeds.

The locusts, too, feed on emphemeral grasses, and lay their eggs in moist sand. Millions of hoppers emerge and move onward when the food supplies are exhausted. During rainy periods the aromatic chemicals of certain shrubs stimulate a change from hopper to winged phase, and the locust-swarms take off on their long journeys.

For the larger animals the plant growth is associated with reproduction. For gazelles it provides food during the mating season. Interestingly, some rodents build up a food store before mating, which means that in dry years the breeding is restricted.

Fig. 10.28 The deadly Australian brown snake lives in arid country, yet finds sufficient small mammals and reptiles for food. The balance of population is easily tipped, either by climatic change or human interference. As Newman, Western Australia's iron ore town, developed in virgin bush, the countryside suddenly acquired more birds than before, and many more mice; resulting in a huge increase in the snake population, which began to invade the homes.

Fig. 10.29 Bulbous *notocactus*, with ridges on its short, swollen green stems, whose thorns are modified shoots.

The heat itself invokes many responses. Most small creatures are nocturnal, retreating to cracks and burrows by day. Insects and arachnids are mainly active at night. The burrowers can find comfort a metre or so down in loose material; for air is a poor conductor of heat.

The mammals have a relatively small surface-to-volume ratio and cool as their sweat evaporates. Hairs trap insulating air. The desert fox and Saharan hare have large ears which give a greater cooling surface. Kangaroo rats and others cool the air they exhale, and so obtain water when hot, moist air condenses in the cooler nasal passage.

Desert habitats each support particular communities. In Arizona holes in the giant cacti are occupied by woodpeckers, owls and flycatchers, and a species of wren lives in smaller cacti. Most of these are predators, living on snakes, lizards, small mammals and insects.

People battle, often unsuccessfully, with arid and semi-arid environments (p.152); though small groups with age-old experience, like the Australian aboriginals and Kalahari bushmen developed skills which enabled them to live a semi-nomadic food-gathering and hunting life, until modern intrusions interfered.

10.7   The warm temperate winter rain (Mediterranean) biome

Fig. 10.30 clearly shows how the dominance of the sub-tropical high pressure during the summer months, and the incursions of moist westerly air, with occasional frontal disturbances during the winter produce a marked summer soil-water deficit, and a rapidly declining surplus from mid-winter on. The present vegetation in these warm temperate lands is dominated in most areas by plants with small leathery leaves, which cut down transpiration by closing stomata when water is scarce (**sclerophyllous species**). Many others are adapted to the deficit, such as cistus and thyme, whose soft leaves wither during summer, and may be shed in extreme drought.

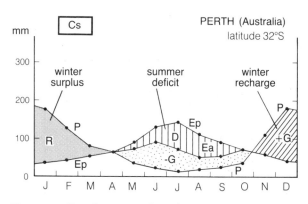

Fig. 10.30   From late winter the soil-water surplus gives way to a deficit throughout the hot summer months. As winter replenishment begins, autumn sees a flush of plant growth, a colourful season in Western Australia's shrubland.

In each of these regions the plant–soil systems have been disturbed by human activities. Around the Mediterranean itself, families and their domestic animals have been clearing and farming the moister lands and grazing the drier hillsides for thousands of years.

The present, unique climate has affected these lands only recently, on the geological time-scale. It is probable that summer rains gave way to summer drought some two million years ago, when plant species growing in drier habitats gradually became the dominants in the vegetation. Species from adjacent drier zones invaded, much as they do when human activities lead to desertification.

Ice advances did not destroy the existing vegetation as they did in higher latitudes, so plant communities have representatives of long-surviving families. The colder, wetter periods brought modifications, especially in mountain areas. In the Mediterranean region the original zonal vegetation, now only found in small remnants, was an **evergreen sclerophyllous forest of trees and shrubs**. Today, in the different conti-

SCLEROPHYLLOUS VEGETATION

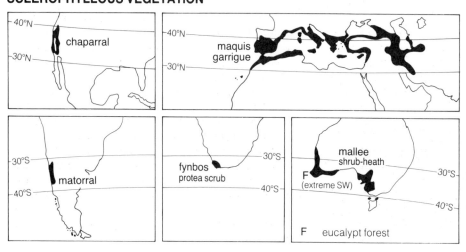

Fig. 10.31   Different species of plants and different names: but each describes associations formed of heath plants, drought resistant herbs, and seasonally appearing geophytes.

nents, each of these regions has its own particular floristic composition, adapted to growth in spring, when the soil is moist and temperatures rising, and in autumn as the rains begin; and above all able to withstand summer drought. Few species can tolerate lengthy cold periods.

The Mediterranean lands themselves are mostly covered with shrubby vegetation, cultivated trees, many of great antiquity; with concentrations of intensive farming. Evergreen oaks, including the cork oak on siliceous soils in the western Mediterranean, and also pines are common throughout, but with favoured habitats: Maritime pines occur on siliceous soils below 500 m; Umbrella pines up to 1000 m; Aleppo pines mainly in the eastern Mediterranean. The grape vines, carobs and olives had been introduced from Asia by the second millenium BC; mulberries, cherries and apricots followed; and citrus fruit from south-east Asia.

Species of shrub from the original forest, such as box, viburnum and rose species form an association with plants like cistus, heather, and broom, generally known as **maquis**. Cistus maquis covers much of the drier, hotter parts of the Mediterranean. In dry locations, and especially

Fig. 10.33 Red soil fills pipes in calcareous limestone in Crete, with garrigue vegetation above.

on limestones and other pervious rocks, lower, rounded shrubs form a more open vegetation – a degraded form of maquis known as **garrigue** – with thyme, sage, lavender and rosemary plants protected by resinous aromatic oils.

The garrigue-covered hillsides are usually a blaze of colour during spring, for there are also geophytes such as tulips, crocuses, irises, garlic, and cyclamen. In very dry areas there are deep-rooted herbaceous perennials and thistle-like plants of the daisy family. Some, like the asphodel, have underground storage organs. Asphodels are regarded as an indicator of impoverished soils.

Some of the dry shrub vegetation is a climax form, due to natural causes; but **much is the result of people's interference**. The contrasts in altitude, maritime influences, shelter, soil characteristics, and human actions give rise to many variations within these broad types of vegetation. Some landscapes are dominated by recent introductions, like the Australian eucalypts and various succulents, such as the American prickly pear.

In **California** woodland also includes evergreen oaks, with some deciduous species. The sclerophyllous shrub formation with heathy plants is known as **chaparral**. Towards the dry south it decreases in height and occurs at increasing altitudes on the hillsides. The shrub *Ceanothus* survived the glacial period and has been introduced to Europe. There has been wholesale destruction of natural vegetation in lowland California.

Central Chile was once covered with dense sclerophyllous shrub associations, with taller trees. Evergreen oaks are absent, but remnants of former woodland peculiar to Chile include the *Quillaja*, of the rose family. Generally, however, the **matorral** resembles the maquis and chaparral of the other regions.

Fig. 10.32 Maquis on a Cyprus hillside, with thick deep-rooted shrubs, broom in flower, and conifers on the upper slopes.

Fig. 10.34 South Africa's colourful proteas, with thick glossy leaves.

South Africa's Cape Province has large areas of climatic climax sclerophyllous scrub; in some places, however, it has been degraded from woodland. Many of the low plants, with leathery blue-green leaves, belong to families not found in Europe. Species of *Protea* occur as dominants. Most are deep-rooted and have large leaves, with a thick cuticle. They can thrive on acidic soils. As with other species in these zonal biomes, fire tends to stimulate growth. It also allows an abundance of herbs, and geophytes like gladioli, to flourish while the slower-growing protea are re-establishing themselves.

In south-west Australia many of the shrubby plants are of the genus *Eucalyptus*, often in mallee form, with branches from an underground tuberous stem. Many of the brightly flowering geophytes are unique to the country. The proteas, particularly the *Banksia*, are often dominant in sandy areas; where the unusual grass-tree of the lily family (*Xanthorrhoea*) occurs in the mallee heath. The variety of soils, developed on coastal limestones, inland granites, and iron-rich duricrusts, creates variations in plant associations.

The eucalypt species also change with the rainfall. In the wetter areas of the south-west are forests of tall jarrah and karri, with trees up to 85 m tall. Huge areas of scrubland have been cleared for wheat and sheep farming, but it readily reverts to mallee when neglected or abandoned.

In the winter rain areas of South Australia and north-east Victoria, this type of vegetation tends to provide the least favourable-land for extensive farming; and occurs only on the fringes of the vineyards and areas of intensive land use, which depend heavily on irrigation. Here the rainfall is erratic, but despite lengthy drought, a rainy period causes a surprisingly rapid recovery of dormant species.

Fire and its effects

The fires which frequently sweep through these forms of vegetation are, in part, beneficial. In the chaparral few sprouting shrubs are killed. New shoots from survivors eventually form thick bushes. These gradually shade out the abundant herbaceous growth which occurs after the fire. After about five years the vegetation is often thicker than before.

The Australian bush fires are usually very fierce, due both to the accumulation of dry litter and the vaporisation of leaf oils from the sclerophylls. Nutrients released from the litter stimulate new growth. We have seen that the grass-trees bloom only after fire and *Banksia* species do not regenerate until fire causes the woody fruits to open. Many seeds are resistant to heat. After fire eucalypts tend to sprout from buds beneath the bark. As we have noted, in South Africa this form of flora also survives fires and benefits from nutrients released from the ashes.

Ecological adjustments are often complex. In California the broad-leaved herbs are the first to recover from burning; they provide extra grazing for deer, whose reproductive rate rises for a short period after fire. The larger deer population then has to cope with overgrazing, which can lead to soil erosion.

10.8   Warm temperate humid biomes

These forests of broadleaf evergreens, with epiphytes and lianas, mostly lie on **continental east coasts**, exposed to the warm maritime air of the sub-tropics. Rainfall is plentiful, usually some 1200–3000 mm, and well-distributed through the year, though almost all have a marked summer maximum. Fig. 10.36 shows that a brief soil-water

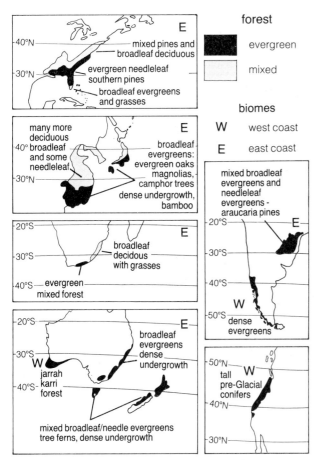

Fig. 10.35 The warmth and humidity favour the growth of forest, usually with dense undergrowth; though the plant species vary from region to region, and with altitude, and topography.

deficit during the warm winter is quickly made up. Deciduous characteristics are most marked in the higher latitudes.

These forests lie in **a transitional zone** between the lower sub-tropics and the cool temperate lands, so there is **a mosaic of different plant associations**, which vary with the soils: for instance, in south-eastern USA, amid broadleaf evergreen, extensive stands of conifers – pines, cedars, and cypresses – occur on sandy soils.

Fig. 10.36 Precipitation is slight during mid-winter, but the brief soil-water deficit is quickly made up. Plant growth can continue through the warm winter with active transpiration.

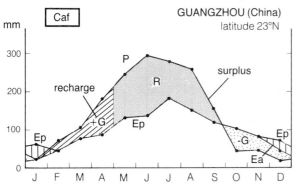

In North America and in Asia, especially, the more northerly parts are affected by very cold winter winds. The forests are mixed; partly in the sense of a patchwork made up of broadleaf associations and separate coniferous associations, and partly because broadleaf evergreens and broadleaf deciduous trees intermingle.

Over the warm, moist lowlands of south-eastern USA the woods include sweet chestnuts, evergreen oaks, magnolias, and a dense shrubby undergrowth. There are numerous climbers, and epiphytic 'southern moss' festoons tree branches. On hillsides between 550–750 m beech joins the associations of plants. Sandy areas and swamplands bear species of conifer, and in the Florida swamps small bush-covered islands with cypresses and palms stand above shallow water thick with rushes and sedges.

In south-east China and Japan evergreen oaks, camphor trees, and magnolias are often draped with climbers, part of a dense cover of smaller trees and shrubs which includes bamboo thickets.

In central China conifers appear among broadleaf evergreens, and there are deciduous species further north. However, with such extensive clearing and large-scale replanting over thousands of years, descriptions of a typical natural zonal vegetation are unrealistic.

Fig. 10.37 Delicate, branching bamboo fronds among evergreen forest trees on the bank of the Lancang jiang (Mekong) in southern China.

In south-eastern Australia the dominants are tall eucalypts, with dense tree ferns among the lower layers. In western Tasmania and the **western coastlands of New Zealand** tree ferns, shrubs and climbers form a thick undergrowth between the tall evergreens. In south-west New Zealand the evergreen beech has an almost impenetrable undergrowth in some areas. New Zealand's forests include the much exploited kauri, and the native softwoods rimu, kahikatea, and totara.

Fig. 10.38 Tall evergreens with occasional palms cover large parts of the tableland of southern Brazil. Trees and shrubs root themselves in the near-vertical face of this valley cut into the basalt rock. Beyond, spray from the Iguaçu falls rises above the forest.

Dense evergreen forest covers the southern parts of **the Brazilian highlands**, its composition modified by altitude. The main dominant is auracaria pine; but many broadleaf evergreen trees and shrubs are associated with it. By contrast, the temperate rainforests of **South Africa** are restricted to a coastal area about 100 km long.

The forests of the warm temperate west coasts are in a different category. **In Western Australia** the jarrah and karri tower above shrub and fern layers. In the much wetter, luxuriant forests of **south-central Chile** there are araucaria pine, evergreen beech, larch, and Chilean cedar.

In northern California, extending northward to southern Canada, are winter rainfall maximum forests with conifers which survived the glacial periods of the Pleistocene. The long-living Redwood *Sequoia* trees are up to 100 m tall.

10.9   The cool temperate humid biome

Before the period of intensive land use, most, but not all, of the lowlands of western Europe, north of the Mediterranean and south of the boreal forests, was covered with deciduous or mixed forest. Minor climatic variations, differences of altitude and slope, parent rocks, drainage conditions, and soils affected the composition of the vegetation locally. Recolonisation by stages after the retreat of the ice sheets resulted in many different woodland associations; as did the variety parent materials and texture of the glacial drift.

In general, the mature soils that developed beneath these forests are podzolic, but rich in organic matter derived from the annual leaf-fall. Immature sandy soils, which are more heavily leached, may support conifers or heathy forms of vegetation, rather than deciduous hardwoods.

But **in Europe there are virtually no virgin deciduous forests**. The structure of the present forests is determined by how people have used them, and how they now manage them. As Figs. 10.40 and 10.41 show, there are sufficient differences in soil moisture content in summer, even between western and eastern Britain, for forests to have different compositions, and for areas of afforestation and agricultural land use to require different drainage or irrigation practices.

Fig. 10.39   The general distribution of broad-leaf deciduous trees in the northern hemisphere, usually mingled with needle-leaf species. Most of these natural woodlands have been modified by human activities.

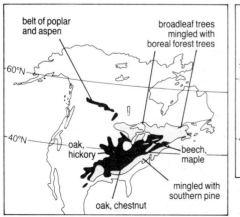

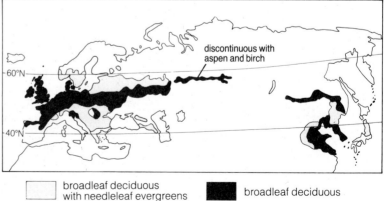

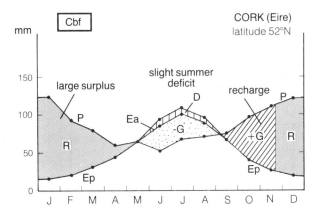

Fig. 10.40 For most of the year there is a large soil-water surplus, and only a small shortage during mid-summer. Perennials continue to grow slowly during the mild winter.

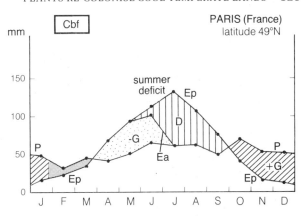

Fig. 10.41 Also in a *Cbf* region; but there is a much greater soil-water deficit by mid-summer; many crops in northern France are irrigated during the period of maximum growth.

In North America the deciduous forests have also suffered from interference; though there are extensive areas of relatively undisturbed woodland, and also many mixed forests. Oaks and hickory occur widely in the woodlands; beech and maple are dominants in the north-east; and oaks, chestnuts, tulip trees, and hemlock are dominants further south; all associated with smaller trees, shrubs and herbs.

The lowland deciduous forest of north-east China and western Korea also contain oaks, beech, ash, and birch. Here, as in Japan, they are co-dominant with conifers; areas of coniferous forest mingle with them, or are separated by altitude zoning. In the southern hemisphere there are thick evergreen forests in the cool, wet coastlands of southern Chile, with deciduous woodland only in very sheltered locations. New Zealand's evergreen southern beech forests, dense with epiphytes and climbers, are found in the cool south-west of South Island.

Deciduous forest ecosystems

The **multi-layered plant community** may have several tree strata, standing above shrubs, with herbs beneath; and geophytes make a seasonal appearance. Trees associated with particular dominants vary with conditions, so that oak-woods include ash on drier soils but birch and hazel on moister ones.

Solar radiation is intercepted by the tree canopy; only a tiny proportion penetrates to the ground when trees are in leaf. The summer temperature of the canopy is much higher than that of the forest floor. The dome-shaped crowns also intercept about one tenth of the rain; the rest drips through or runs down the trunks. In summer there is intense transpiration loss from the crown, and even the herbaceous layer loses about one fifth of this amount; so that the forest uses nearly all the precipitation received.

The deciduous behaviour, with consequent effects of leaf-shedding on the lower layers of vegetation, is linked both to temperature control over the availability of soil solutions and to **light**

Fig. 10.42 During summer the overlapping, domed tree crowns on this Devon hillside intercept almost all the insolation. There is a huge transpiration loss from this upper leafy cover. Beneath, the woodland floor is deeply shaded and cool.

intensity. Leaves which are fully exposed to the sun are generally smaller, thicker, and have more stomata on the underside. Photosynthesis increases in proportion to light intensity, until a maximum is reached. For sun-lit leaves this is about two-fifths of the incoming light energy; but only about one-fifth is required for leaves in the shade, which are better able to use lower light intensities. The upper leaves tend to be steeply inclined to reduce water loss; so that more light penetrates to the lower leaves, which are at right angles to the incoming light.

Fig. 10.43 Here, where light penetrates, is a close undergrowth of shrubs, young trees and ferns.

When the illumination is below a certain minimum, photosynthesis does not match up to respiration; so, to avoid losing plant mass, the leaves turn colour and are shed. The required light values vary with tree species.

Transpiring leaves are also shed in response to the inability of the tree to absorb and circulate soil minerals as the temperature falls in autumn. The damage to tissues due to freezing and the drying out of aerial organs at low temperatures also control the distribution of species; some of the hardier being able to resist very cold conditions through changes in the cell protoplasm.

Light conditions also influence competition between light-demanding and shade-tolerant trees, and seasonal changes in light intensity greatly affect the vegetation of the forest floor. During summer the light intensity at ground level is low by comparison with that of the crown; the temperature is more moderate and the humidity higher, which favour shade-tolerant, hygrophytic plants. Most of the intense radiation comes only in flecks of light, as the branches move. Tree seedlings, and plants like wood sorrel reduce respiration as leaves cut down the light that reaches them.

During spring, when the trees are still bare and the light intensity is increasing, **the forest floor** is adequately illuminated. Geophytes, such as bluebells, scilla and anemone, benefit from the light, and from the heat of the litter layer. During this brief period they flower, fruit, and store energy reserves in their underground organs. Most species of the herbaceous layer have buds at the base of the shoot, protected by leaves and snow during winter. The herbs themselves form a litter which is rapidly mineralised, compared with the tree litter, and help to recycle nutrients. They are, however, in competition with tree roots, which are better able to obtain water from the soil. As a result of this, the floor of beech forests in particular has few herbs, though the slow recycling of the thick leaf litter is another factor.

In most deciduous forests, however, **the litter produced is annually recycled**, and its removal can lead to the impoverishment and acidity of forest soils, and a decrease in the mass of wood. The productivity of the forests varies with the dominant species and plant associations, and with the ratio of the leaf area of the tree to the ground area covered by it. **The mean primary productivity** of deciduous forest is some 13 tonnes of biomass per hectare per year, compared

Fig. 10.44 Different forms of vegetation maintained by careful management. These upper slopes on Exmoor were cleared of woodland long ago, and bear heath vegetation. Rows of young conifers have now been planted on the lower slopes. Deciduous trees have been established for shelter, as hedge plants, and along the valley. Below the house there are fields of pasture grasses, but above this bracken is successfully advancing over the steeper sown grassland.

with a mean of 23 tonnes for tropical rainforest, and about 8 for northern coniferous forests.

Small animals, such as earthworms and insect larvae, consume the litter and break it down, making it more accessible to micro-organisms, fungi and bacteria, which, in any case, immediately attack the dead organic material. Litter mineralisation continues in the O layer of the soils, where animal organisms are present. But there is not the immediate, complete cycling between plants and litter nutrients that takes place in the tropical rainforests.

Fig. 9.2 shows a simplified **food chain in a deciduous woodland**. Links in the chain have important functions for the whole community at different levels. Bacteria and fungi are vital components at the litter level. Various rodents cause damage by nibbling tree seedlings, consuming the herbaceous layer, and collecting food hoards which are not completely used up; but their subterranean passages and the soil heaps removed improve aeration and provide a considerable volume of earth, rich in mixed nutrients.

Here, too, relatively undisturbed secondary woodland can acquire a balance between the different life-forms, and the thoughtless removal of even one species by human actions can affect the whole community. The destruction of the habitat of owls, which has resulted in a steep, progressive decline in the owl population in Britain in recent years, also affects the population of the creatures on which they prey. The numbers of small rodents may greatly increase, with repercussions through the entire food-chain.

Other sub-climax vegetation

Grasslands, maintained at a sub-climax stage by pasturing and mowing, form much of the landscape in the moist cool temperate lands. Cropping tends to form a compact turf, encouraging the growth of axillary buds and shoots and the development of the fibrous roots. Today this is a maintained vegetation. Neutral soils carry perennial rye grass, white clover, cocksfoot, meadow foxtail, and rough-stalked meadow grass; acidic soils bear fescues if well drained and species of *Agrostis* where not. On wet uplands there may be only purple moor grass with bog plants. Their use as pasture plants can be carefully regulated.

In north-western Europe **heaths** and associated plants occupy dry, loose soils over sands and gravels in lowland areas, where ling is dominant. They also tend to form the vegetation on high moorlands, where rocks, like granite, break down to a thin acidic soil, and moist conditions produce damp peaty soils. Here are ling, and heaths like bell heather, with bilberries, cranberries, and cotton grass. There is *sphagnum* moss in **boggy areas**. But on drier land mat grass and purple moor grass form tussocks, with occasional taller shrubs, such

as gorse and juniper. Many of Europe's heath-covered uplands were once forest-covered. Their recolonisation by tree seedlings has been prevented by grazing and burning. Where the soil has deteriorated, bracken has often taken over.

Many of these plants dominate areas where the soils are not only poor in nutrients but may be moisture-deficient during summer. The leaves of ling, crowberry, and bell heather are rolled, with their lower surfaces reduced in area, and with few, sunken stomata. But the rapidity with which other plants will take over is emphasised by the fact that grasses and grass-heath will replace ling and bell heaths where droppings of sheep or rabbits are plentiful. Sown pastures, improved by fertilisers, can be established very quickly.

Many **lowland moors** are stages in a succession which has followed the drainage or infilling of bog-land. Heaths have taken over from mosses, and already may be associated with scrubby bushes, silver birches, or conifers. Marshes may have progressed to damp elder woodland, along the lines of the hydrosere described on p.107.

Shore vegetation

Vegetation occurs in zones roughly parallel with the shore. These extend from the limits of photosynthetic phytoplankton beneath sea-level to that of the shingle banks, sandy dunes and salt marshes, where colonisation by higher plants may occur, and successions develop.

Apart from free-moving plankton, feeding fish and crustacea, most shoreline plant life is made up of various algae, whose nutrients come from sea-water. They are attached to rocks and boulders **beneath and just above, water-level**. Their form depends on light intensity, temperature, salinity, water movements (tides and currents), and, of course, on competition. Many species change in form from summer to winter.

Girdles of different red and green algae, and large kelps, extend from below tide-levels, where they are exposed by particularly low tides. Above these, brown algae are more common. They are regularly covered and un-covered by the sea. Some, such as bladder wrack, are large; accompanied by delicate eel-grasses attached to the rocks. Where the shore receives only spray, or occasional high tides, the rocks are often covered by black lichens.

The shingle banks bear plants with long tap-roots, usually with flat rosettes and swollen foliage. On **dry sand banks**, salty from spray, there are halophytic plants, like sea-thrift. Marram grass, with long rhizomes and an abundant, spreading root system helps to bind the sand. If covered by sand, fresh roots develop higher up its stiff stems. The leaves are deeply corrugated, and open out fully only in moist atmospheric conditions.

Fig. 10.45 At Kaikoura, New Zealand, coastal uplift has separated the algae vegetation of the tidal zone from that of a low platform, where a dense cover of weeds and lichens is enriched by sea-bird droppings. Above this, the higher raised beach is covered with deep-rooted thistly plants and coarse grasses. Tussock grasses grow on the screes, and small bushes are rooted on the more stable slopes.

Once **the sand dunes** are held, they may be colonised by mosses or finer grasses. Eventually shrubs and trees may enter the plant succession.

Flats, marshes and tidal estuaries develop food chains which exhibit a high productivity. The tide transports wastes and nutrients. Mussels, worms, and other detritus feeders make such efficient use of the bacterial breakdown of organic waste to increase their biomass that commercial management of beds of clams, cockles, mussels and oysters gives a much higher yield per hectare than beef cattle on grassland. Birds, of course, find rich feeding grounds within this ecosystem.

10.10 Temperate grasslands of the continental interiors

Vegetation associations of grasses and other herbs extend over much of the interior of the northern continents, thriving in drier conditions, with greater temperature ranges, than in the maritime lands which favour the deciduous forests. They shade into, or are sometimes interspersed with xerophytic and halophytic forms of vegetation of semi-arid lands. There are transitional zones, too, in the moister areas where grassland and stretches of forest intermingle.

In America these grasslands go by the name of **prairie** (Fr., meadow), and in Euro-Asia **steppe** (where the genus *Stipa* is common). Much of the area formerly covered by wild vegetation is now either cultivated or used for controlled grazing. There are temperate grasslands, too, in the southern hemisphere: the Argentinian **pampa** (Sp., grassland); the South African **veld** (Dutch, grass); and tussock grassland in New Zealand. But in these southern hemisphere areas climatic controls and human influences are somewhat different, so we shall look at them separately.

Many of the temperate grassland soils in the northern hemisphere have developed on transported material. Particles carried by wind, water, or ice have been deposited, and have accumulated, over the plains. Minute dust particles from arid lands have settled and built up a loess cover of considerable depth in some areas; those in eastern Asia are immensely thick. Large parts of

Fig. 10.46 The temperate grasslands of the northern hemisphere. The nature of the vegetation, as Fig. 10.47 shows, responds to the increasingly arid conditions inland.

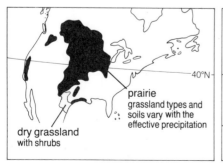

prairie
grassland types and soils vary with the effective precipitation

dry grassland with shrubs

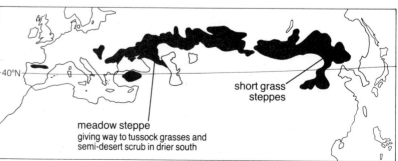

short grass steppes

meadow steppe
giving way to tussock grasses and semi-desert scrub in drier south

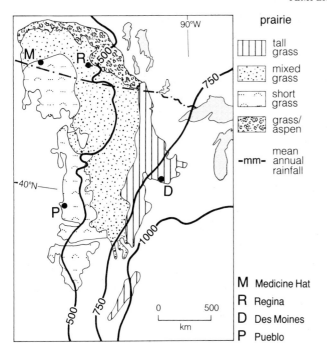

Fig. 10.47   The plant associations on the prairies change as precipitation decreases and altitude increases westward. The latitudinal variations in insolation are also significant.

the North American prairies are mantled with glacial deposits, much of which has been redistributed by outwash. Drying glacial material has been a source of a loess covering in neighbouring areas.

The mixture of minerals in each of these deposits provides a wide range of nutrients for soils of high potential fertility – though there are also infertile areas of outwash gravel and sand. But sufficient moisture must be present to make such nutrients available for absorption by plants. Fig. 10.47 shows the relationships between annual precipitation and grassland distributions in the prairies.

It is important to realise that there are many thousands of grass species, each capable of tolerating specific ecological conditions, and that a variety of other herbaceous plants, with varying tolerances, make up the grassland vegetation. Thus around Des Moines (Fig. 10.47) the tall-grass prairie is floristically rich, abundant with herbs, and bright with flowers in the moist mid-summer months. The big blue-stem grasses are 1–2 m tall and have a dense root network. Here the rich soils remain moist to a considerable depth. But around Regina the dominants are grama, wheatgrass and buffalo grass, all short sward-forming grasses. Here the low temperature and dryness during winter enforce long rest periods. In the short-grass prairie near Pueblo on the higher, drier western plains there are fewer species of herbaceous plants. In any case, grazing now kills off grasses of middle height; and in each of these regions grazing, cultivation and fire have modified the former associations.

The grasslands are an open system, where ready adjustments to interference, as well as local soil variations, create **a mosaic of different plant associations**. But **cultivation reduces floristic abundance to just a few species**.

Grassland composition

The grass family dominates these ecosystems. In most species the blade of the leaf is nearly vertical, giving a high ratio of leaf surface to ground area, and so efficient productivity through photosynthesis.

The grass stem increases in length from a small region at the base and just above each node. Grass thus stands up to grazing, cutting, and burning better than most plants; in fact such activities stimulate growth by encouraging side shoots from axial buds when the main shoot is removed.

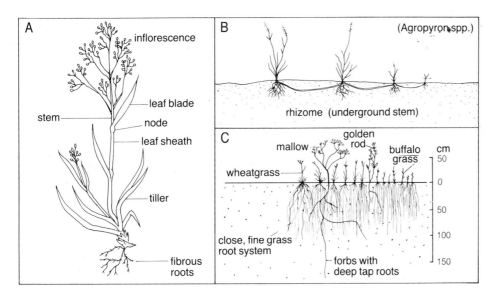

Fig. 10.48   Prairie grasses, forbs, and their root systems.

Many perennial grasses have underground stems (rhizomes). Those with close nodes, from which new shoots emerge, are bunched together in a tussock form. Others have long creeping rhizomes with nodes more widely spaced, from which new shoots and a large volume of thin, fibrous branching roots develops, forming a matted mass, or turf. The root hairs absorb water and soluble nutrients.

The long, flowering stems produce seeds, distributed mainly by the wind or by animals. They then remain dormant, sometimes for long periods, until conditions favour germination.

Besides the grasses there are other herbs, sedges, forbs, legumes and geophytes, though a dense grass mat tends to prevent the development of other plants. The appearance is continuously changing as the grasses mature and wither, and bulbs such as iris, herbs like the salvias, and the various daisies (*compositae*) flower and die down.

All these create a great amount of litter, whose decomposition releases potential nutrients to the soil. In the moister regions the many worms and other small fauna mix soil particles and nutrients, and create spaces which contain oxygen required by bacteria and other decomposers. **Organic decomposition** is at its highest in spring and early summer; though even then old organic material remains and helps to form a thick matted sod, whose mass may exceed that of the plants above the ground.

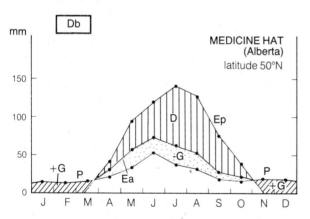

Fig. 10.49 After the great need for water during the summer, the Ep of the short grass prairie falls to zero during the severe winter. Snow accumulation becomes available to recharge the system during the spring melt, and helps to maintain a soil-water storage despite the summer withdrawals.

The plants and the matted layer intercept a high proportion of rainfall. Percolation into the soil is restricted. As transpiration rates are high in summer, and as growing plants take up solutions, this may create a water deficit close to the surface. But the close mat does help to protect the surface particles from wind erosion and sheetwash.

The precipitation amounts vary considerably, especially in the drier parts. In western Kansas,

over a 76-year period only 12 per cent of the years came *near* the statistical mean. Here during wet years the tall grasses increase to form an upper storey over the short grasses. But during long droughts, even the ground cover of the short grasses is reduced from some 80–90 per cent to 20 per cent. Grasshoppers increase in dry conditions and help to eliminate herbaceous species; and xerophytic plants, including cacti, are likely to spread.

Prairie animals and grassland ecology

As we have seen, animals in the soil help to maintain its texture and mix nutrients. The open grassland affords little cover for the bigger animals; so the large herbivores have always herded together for protection, and have continuously migrated to avoid over-using the forage. Compared with the savannas, there are **few dominant species of large mammals**. The most common are the buffalo and pronghorn antelope of North America, and the wild horses and saiga antelope of the Asian grasslands.

The prairies once supported tens of millions of bison and pronghorn on the drier grassland, and elk near the grass–woodland margins. Their grazing, hoof impacts, and droppings affected the ecological balance of the grasslands, as did the activities of other animals. When the North American Indians began to use horses introduced by Spaniards to hunt buffalo, the impact on the herds was slight, in view of the small tribal populations. Their fires had great effects on the grassland, and helped to prevent forest regenerating in marginal regions. But over the decades, as white settlers pressed westward in the late 19th century, the great majority of these vast herds was destroyed.

Today, it is the **burrowing rodents** which particularly affect the plants and soils. Ground squirrels, gophers, mice and kangaroo rats all bring enormous amounts of material to the surface. Prairie dogs dig 3–4 metres deep, and live in territorial colonies. Badgers are also widespread, and dig for rodents.

Jack rabbits use tall grasses for shelter, for predators like the coyote live in open grassland, but feed off the short grasses. There is a complicated food web. Mice, voles, and ground squirrels eat seeds and bulbs and also the abundant grasshoppers, crickets, beetles and spiders. The biomass of the insects is huge. They are consumed by larks, sparrows, pippits, prairie chicken and other birds. Snakes feed on lizards, frogs, toads, worms, and some on squirrels and voles.

In open systems like these, fluctuations in one animal population reverberates through the others. A decrease in the number of predators leads to an increase in the burrowers, with a consequent deterioration of the grassland surface and a decrease in smaller life-forms on which they

Fig. 10.50 Landscaping ridges and contoured ploughing attempt to prevent a system of gulleys extending headwards across wheatlands in Australia's Murray basin.

prey. People have initiated such a chain by killing wolves and coyotes, leaving myriads of rodents to affect soils and crops.

The removal of the natural grass cover has exposed most prairie soils to the threat of erosion. It was the apparent soil fertility, built up over the centuries, but fragile when disturbed, that tempted settlers to extend wheat-growing to dry lands, too far west. Drought years brought disasters in the 1890s, which went unheeded until the devastation during the long droughts of the 1930s; then countless millions of tonnes of once-fertile top-soil were lost from the great western dust-bowl.

The Eurasian steppes

These extend from eastern Europe far across central Asia, but have less latitudinal spread than the prairies. Aridity increases from the north-west to the south-east. In the cooler, moister west and north grasses and tree seedlings compete. There the meadow steppe of mat-forming grasses resembled the tall-grass prairie. Southwards and east-

wards the grassland is dominated by tufted *Stipa*, with other shorter turf-forming grasses.

Towards the interior conditions become drier and the temperature range greater. Turf-forming grasses give way to tussock forms, with xerophytic shrubs in the semi-arid lands. Far away to the east are the short-grass steppes of Mongolia. In the west, where the chernozems are cultivated, little natural vegetation remains. As on the prairies, over-use of the drier brown soils, further inland, have led to serious wind erosion. Here, too, people have completely exterminated big game and predators, leaving vast numbers of rodents. Insects which were harmless in a balanced steppe ecosystem now attack grain and sugar beet in large numbers.

10.11   Temperate grasslands of the southern hemisphere

These are mostly warm temperate grasslands, with less cold winters than those of the northern hemisphere, though cooler tussock lands in Patagonia and New Zealand lie further south.

The pampas of Argentina and Uruguay

These border on the Atlantic coast, with an annual rainfall of some 1000 mm in the north-east, falling to 500 mm in the dry south-west. But summer temperatures and the potential evaporation are high. Only in the coastal regions of the river Plate are the annual rainfall and evaporation roughly equal. In the more arid parts evaporation exceeds precipitation by as much as 700 mm; by mid-summer the vegetation is parched.

Climatically these conditions tend to favour grassland. So does the fact that parts of the pampas were burnt over, even before the period of major European settlement. Yet trees grow well; and form shelter belts about the homesteads of the large estancias. Almost nothing remains of the original vegetation, however. European grasses and alfalfa have been introduced to suit the European cattle breeds, and grains are grown commercially. But the soils indicate that there were few trees here before.

Fig. 10.51 Grasslands of the southern hemisphere.

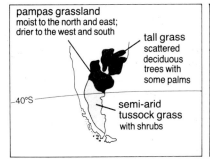
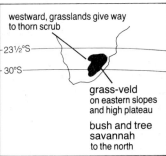
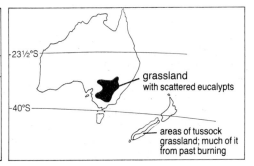

In the dry south-western parts the natural vegetation is mainly tussock grass, with few herbs: and in most of the eastern parts of Patagonia it consists of low tussock grasses and xerophytic cushion plants; an adaptation to the persistently strong winds.

In the more rolling countrysides of Uruguay and southern Brazil, woodland occupied depressions, and still covers parts of the valleys. Occasional palms stand out above the grasslands.

New Zealand tussock country

In the eastern parts of South Island, in the lee of the Southern Alps, low tussock grasses predominate. The rainfall is only some 300 mm; and here, woodland covering the eastern hill country and plains was fired centuries ago by the Moriori and Maori people. Since then grazing has prevented seedlings regenerating a woody vegetation. In other parts of New Zealand much of the hill country cleared of forest during the last 100 years has degenerated to poor shrubland with coarse grasses.

South African grasslands

Here the temperate grassland, or veld, is a natural climax vegetation above 1400 m in the eastern interior. Frost and winter drought have favoured grasses rather than trees. On the highveld the grass is short and trees only appear in galleries along rivers. In the north, however, in the lowveld, this gives way to taller grassland and bushveld, which is a type of savanna, with acacia predominating. Here fire and overstocking have created open, shrubby, tall grass country in places. Towards the drier west it becomes a dry, thorny shrub savanna.

South-east Australian grassland with trees

Stretching away westward from the open eucalypt forest of the inland slopes of the eastern highlands is dry grassland with scattered eucalypts, and galleries of tall red gums along the seasonal watercourses and rivers. Westward this gradually gives way to mallee eucalypts, tussocky grass and saltbush, or to tracts of arid mulga scrub.

10.12   Vegetation of the middle latitude arid and semi-arid lands

Most of these arid lands lie far into the interior of the continents, at varying altitudes (Fig. 5.4). Many are extensions of the drier parts of the temperate grasslands, which **in North America** merge westward into sage-brush country, with low, drab-coloured shrubs. As conditions become drier, these give way to patches of wiry grass, creosote bush and woody stemmed shrubs, some deep-rooted, some with wide-spreading root systems. The plants and their responses to aridity resemble those described on p.121. Cacti, euphorbia and other succulents are common, and in the dry valleys deep-rooted acacia and mesquite bushes make use of the relatively higher watertable.

In central Asia precipitation diminishes from west to east. In the southern parts occasional cyclonic disturbances from the Atlantic bring rain or snow during winter. Further north the rain is mostly in spring and summer. Evaporation is low during winter; but in the Karakum desert, east of the Caspian Sea, the hot summer is a time of extreme drought, for the potential evaporation is some 10–15 times the precipitation. Much of this is a sandy dune desert; but wide expanses of dune are plant-covered, emphasising **the extent to which sands can hold water in really arid locations**.

Fig. 10.52 Among the dominant plants of the middle latitude deserts in the USA are well-spaced, deep-rooted shrubs, such as sagebrush, and taller succulents like yucca and cacti.

The upper 2 m of sand contains available water, so that shrubs remain active through the year. On the lower dune slopes there are spring and summer ephemerals; and in deep dune valleys, where the water-table is at a depth of over 5 m there are low, woody, salt-tolerant trees. The vegetation was once sufficient to support antelopes and wild horses. Today, millions of karakul sheep graze the sandy desert. They tread seeds from the shrubs into soft ground, which facilitates germination, and spread their droppings over a surface which is churned up by innumerable rodents. Giant tortoises feed on ephemerals before hibernating.

In the valleys of rivers fed from distant snow-fields, poplars, willows, elms, and ash grow amid grassy swards in closely settled green oases. By contrast, many parts of central Asia remain barren and rock-strewn, almost devoid of vegetation, for there is little precipitation and extremes of cold and heat.

In central Asia, as in North America, **inland drainage areas** contain salt lakes and salt flats with occasional halophytic plants. But over such vast distances there are innumerable variations in plant associations, with alpine plants influenced by altitude. In the north the short summer following the spring snow-melt allows the moist ground to support larch and a carpet of low shrubs.

10.13   The boreal forest zone

These **largely coniferous forests** extend across huge areas of Euro-Asia, where they are known as **taiga**. The southern limits are about 50°N in the east and 60°N in the west. In eastern North America they extend as far south as 45°N. There are considerable variations in the composition of the forest between the milder maritime areas, such as Scandinavia, and the centre of Asia, which experiences temperatures of +30°C and −70°C. But near their northern limits the forests fade into, or intermingle with tundra. In general, they replace broadleaf deciduous species where the summers are too short for them (less than _ days above 10°C), and the winters dark and col((below 6°C for more than six months). Winter precipitation is mainly as snow, of course, and though the amounts are small in the continental interiors, it tends to accumulate.

Conifers are the main dominants: in Europe they are mainly spruce, particularly in moist soils, with pine in drier areas; in North America there are rather more species. They are able to maintain themselves under the normal hard winter conditions by virtue of their narrow needle, or scale-like, leaves, whose cuticles give strength in times of water deficiency and enable them to reduce transpiration. In central Asia, with such extremes of cold, only small deciduous birch, larch, and aspen may be able to survive.

The evergreen conifers retain their leaves during winter, even when soil water is unavailable; so photosynthesis can take place as soon as conditions allow. The individual leaves usually last for a number of years, though larches shed theirs in autumn, and, with deciduous birch and aspens, can survive in colder locations than the evergreens.

The conifers absorb few bases and can thrive on acidic podzols which do not suit most broadleaf trees. Strands of fungi extend their root system and make organic nutrients available. The roots are shallow and can flourish where the sub-soil remains frozen for long periods. The trees are softwooded, with thick resinous bark and shortish springy branches, which help to shed snow. Their overall, compact shape also helps to prevent snow accumulation. The more severe the climate, the narrower the tree; for the terminal growth continues after the side twigs have ceased to develop. They stand well against the wind; though, being shallow-rooted, exceptional gales are apt to up-root them.

In the cones the ovules are on the cone-scales, and protected when these close together. Under favourable conditions they open out to free the seeds.

Fig. 10.53   Coniferous forests of the northern hemisphere.

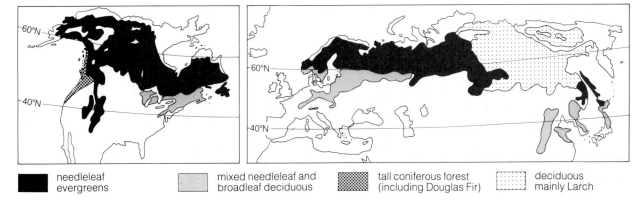

| needleleaf evergreens | mixed needleleaf and broadleaf deciduous | tall coniferous forest (including Douglas Fir) | deciduous mainly Larch |

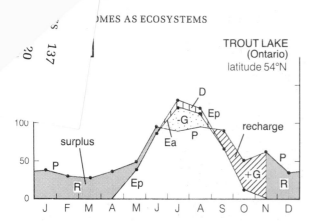

Fig. 10.54 The water need is virtually nil during the bitter winter months, but rises sharply during the short summer, when water is withdrawn from storage. But first snowmelt and then precipitation recharge the storage, which remains high.

The forests are usually of continuous stands of dominant tree species. Spruce litter does not decompose readily, making it difficult for other seedlings to establish themselves. The floor of the forest is often mossy, though a herb layer with bilberries, or cranberries, and wood sorrel grows in the drier locations. Pines tend to displace spruce in dry habitats and are associated with heather and the kind of herbs mentioned.

Summer fires can generate great heat and cover large areas. Herbs and small shrubs quickly re-colonise burnt sites, while birch and aspen are usually the first trees to establish themselves, until ousted by growing pines, while the spruce develops more slowly beneath.

Amid the Canadian spruce-fir forests, moose and caribou are browsers, using the shelter of the trees, and able to live on mosses and lichens as well as the grasses, herbs, and tree seedlings during summer. The animals destroy much of the local sub-climax vegetation, and so help to prevent fully developed forest taking over. When their numbers become too large, they can damage their food supplies and reduce feeding grounds to mud patches.

Among the insects at the base of **a complex animal food web** are budworm moth caterpillars. These feed on buds during spring, and can defoliate and kill trees over a wide area. Trees are also at risk from spruce beetles which make galleries under the bark. A variety of birds, including many species of warbler and woodpecker, act as controls. Many small animals inhabit the ground and fallen trunks, and tree martens feed on squirrels, mice, and voles. Snowshoe rabbits kill trees by removing bark above the snow-line, and periodically their population becomes dangerously large, despite their control by lynx, fox and wolf.

Despite the almost monotonously empty appearance of immense stands of boreal forest, **animal life abounds** within the biome, with continuously shifting balances **in its various ecological niches**. Beaver colonies and their dam-building activities are well known; as are the bears, which as omnivores destroy shrubs and trees in their search for buds, fruits, and berries; and the porcupines, which, as bark-eaters, ring-bark and damage conifers.

It is said that **plantations** are replacing the enormous volumes of softwood obtained daily from the boreal forests for the world's newsprint and other purposes. But, as with the other forests, **it is impossible to regenerate the natural balance** which is destroyed when even a small number of the natural species are removed, let alone millions of hectares of virgin forest.

10.14 The tundra

Tundra literally means a treeless plain. The 10°C isotherm for the warmest month is an approximate dividing line between tundra vegetation and that of the boreal forest; but, of course, there is

Fig. 10.55 Lodgepole pines extend to the timberline, where sunlight stimulates growth above winter inversion levels; but at that height trees cannot cope with water losses when the ground is frozen. Winter avalanches cut swathes through the forest. Any fire damage is soon made good by a flush of growth, for cones release seeds under great heat, and for a while foxtail grasses and fireweeds provide extra grazing for the elk and moose.

Fig. 10.56 A small settlement in east Greenland strewn with glacial erratics on treeless tundra. It is sheltered by a rock face, on which frost shattering has created great screes. Peat is cut and insulates the roof of the stores (**left**).

inter-penetration on either side of that limit, depending on variations of topography, soils, and micro-climates. There is usually a forest–tundra zone where the boundaries between the two advance or retreat with changing conditions. Away from the forest stands, the boreal trees are partly handicapped by the low density of their seed; and, of course, the capacity for germination is much reduced in these habitats, where winters are dark and cold, winds strong, and summers short – though with almost continuous insolation. Above the permafrost the spring-melt makes hollows swampy, and the soils are generally peaty with gley characteristics. Here micro-habitats support quite different associations of small plants.

Overall, **the vegetation forms a close, almost continuous sward**, with hardy grasses and sedges, tiny hygrophilous herbs, and small berry-bearing plants like the crowberry and bilberry, with occasional shrubs, dwarf willows, or dwarf birches. There are numerous mosses and lichens.

Most plants are able morphologically to check transpiration when frost hardens the ground and prevents them absorbing water; though many species *are* adapted to absorb water when the temperature is very low. Even in early summer, drying winds and rising temperatures cause a high rate of evaporation while the roots may still be frozen. But many plants have a cushion form, low to the surface, and take advantage of upper soil temperatures, which at this time of year may be higher than the air temperature.

The growing season is short, but **many hours of unbroken sunlight** each day allow stomata to remain open for long periods, so **there is a very great increase in tundra biomass**.

The plants must complete their development cycle in this short season, so are adapted to commence photosynthesis as early as possible. In early summer the profusion of flowers and the various shades of green produce a brilliant colour-

ing. This is maintained, as many plants soon bear bright berries, like the shiny red myrtle berries, on which so many animals feed. In some species seed germination begins before dispersal, so that the plants may become established in time to survive the hard winter. In others the seeds may not mature for several years after the flower buds have formed.

In both flora and fauna the number of species is relatively small, while the number of individuals in a species is relatively large. There are few annual plants. **Most of the mass is in the woody herbaceous perennials**, with underground storage organs, and small, thick evergreen leaves. These make up heaths which cover large areas in mid-tundra, though variations in the physical habitat make for discontinuity. The rushes and sedges are found in wetter hollows with hardy grasses on better-drained sites. In the mosses and lichens the tissues do not freeze, and they are capable of photosynthesis at temperatures as low as −20°C.

Some areas with nutrient-deficient soils are poorly vegetated, and contrast with sites favoured by birds and large animals, whose droppings allow a close cover of plants otherwise unable to establish themselves.

Micro-variations are extremely important. Variations in drainage create vegetation differences on upheaved surfaces. Coarse material of screes and solifluction slopes may preferentially favour plants with long underground stems. Variations in snow depth, and the duration of exposure of the surface beneath, can give rise to small catenas, where each species occupies a position suited to the length of exposure.

Towards the south, the longer growing season and warmer summers allow deciduous shrubs to develop to some 2 m tall, with thickets of dwarf willow, alder, or birch alternating with herbaceous and marsh plants, according to local conditions. Here and there, to aid reindeer graz-

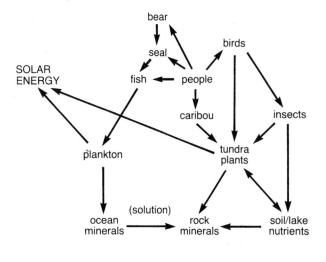

Fig. 10.57   An Arctic food chain.

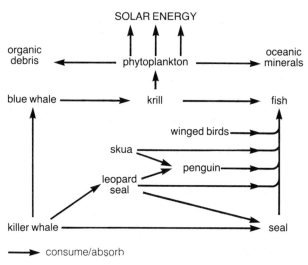

Fig. 10.58   The oceans as food sources.

ing, tree seedlings, vulnerable to competition, are controlled by burning, so extending the tundra.

Food chains tend to be simple. The shorter the chain, the greater the biomass from a given amount of energy. Large animals leave the tundra during winter, and the birds migrate to the south, leaving behind the lemmings and ground squirrels. The former remain active, feeding on small leaves and buds beneath the snow. As they live in colonies, they can destroy much of the local vegetation. Population surges and lack of food cause their occasional mass migrations to self-destruction.

The flocks of water birds, and predators like the polar fox and snowy owl, return with the spring. Reindeer migrate with the seasons; though now domesticated herds far outnumber the wild animals.

Cold-blooded insects are able to survive by hibernation, and the ability of larvae to endure sub-zero temperatures. Most complete their development over two summer seasons. In summer they reproduce in large numbers, and European tundra mosquitoes and Canadian biting black-flies plague travellers in these regions. Nostril flies and warble flies torment the caribou.

10.15   Arctic and Antarctic Ocean food chains

Many of the land animals and birds depend on sea-water minerals and single-celled green phytoplankton as bases of their food chain. The sun's energy acts to synthesise a mass of phytoplankton, which are consumed by vast numbers of tiny zooplankton, among them the minute shrimp-like krill.

The Antarctic Ocean is rich in life-forms. Close to the continent a turn-over of bottom water brings up nutrients released by the decomposition of continuously sinking organic debris. The krill

multiply, build up the body mass of whales in a short food chain, and feed the abundant fish population. Fig. 10.57 shows some of the food chains which link terrestrial plant–soil systems with these marine bio-systems. In the Arctic, land animals, like bears, which eat fish and seals, and the fish-consuming human population all depend on the summer abundance of marine micro-organisms, which in the Arctic Ocean benefit from spores released from the ice.

10.16   Mountain vegetation

In mountain systems there are broad vegetation zones at various altitudes, though the actual plant species respond to the local climatic regime, with its seasonal variations, or lack of them. As the light intensity and hours of daylight vary with latitude, there are corresponding differences in the flora.

In low latitudes the changes with altitude are sometimes seen as mirroring the changes in vegetation from rainforest to arctic tundra; but, in fact, the proportion of species, and the characteristics of the individual plants are not the same. Where mountains rise above tropical rainforest, vegetation on the lower slopes broadly resembles that of surrounding lowland. But sunlight penetrates more easily on slopes, producing a denser undergrowth, and the dominants tend to be less tall. Precipitation usually increases with altitude, and, therefore, so also does the proportion of hygrophilous species; a zone of high humidity and cloud favours branching and leafiness. Tree species of the lowland rainforest mingle with sub-tropical species. All have a mass of foliage, often covered with epiphytic mosses and lichens – a so-called **moss-forest**, with an abundance of ferns and climbers. The trees, usually no higher than 12 m, have contorted trunks and boughs.

High mountain slopes west of the East African plateau have an unreal appearance: on Ruwenzori, for example, fleshy grounsel, lobelia and heaths grow to giant proportions.

In a sense, **great mountains make their own climates** and their own related plant associations. Mountain systems may act as barriers to migrating species, and so restrict the numbers of competitors. There are often unusually dry, sheltered locations. Even in low latitudes, exceptionally arid conditions occur among the central and western Andes: extensive areas of dry grassland with xeromorphic plants are found at 3000–4000 m; and while the eastern slopes carry dense forests, the sides of the deep river canyons cutting through the eastern ranges bear arid scrub.

The actual species at any height depends on the latitude, orientation, and extent of the mountain system. At comparative heights in an east–west system like the Himalayas the associations on the drier western mountains and wetter eastern ranges are quite different. Fig. 10.59 shows the altitude zoning in a mid-latitude location in the eastern Himalayas. The low-level vegetation is a mixture of tropical evergreens and deciduous forest, with sal, bamboo, and tall grasses, and shows pronounced seasonal changes. Above 1500 m the deciduous woods contain many trees found in cool temperate latitudes – oak, magnolias, birch, alder. Higher still there are evergreen conifers and rhododendrons, some as high as 3000 m. Above are grassy swards with many small herbs, and tiny, brightly coloured tundra plants and lichens. Alpine plants flower at nearly 6000 m.

In all Alpine locations the light is intense, if intermittent. It is richer in blue and violet wavelengths, and causes differences in plant form and colour. The rarefied air makes for increased transpiration. Plants tend to be dwarfed, with narrow leaves arranged in rosettes. Many are

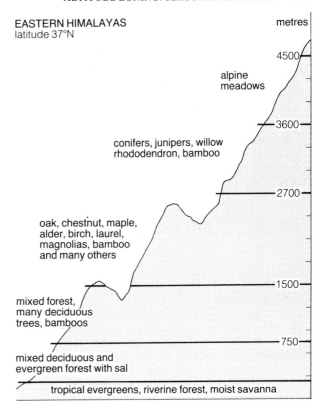

EASTERN HIMALAYAS
latitude 37°N

metres

4500

alpine meadows

3600

conifers, junipers, willow rhododendron, bamboo

2700

oak, chestnut, maple, alder, birch, laurel, magnolias, bamboo and many others

1500

mixed forest, many deciduous trees, bamboos

750

mixed deciduous and evergreen forest with sal

tropical evergreens, riverine forest, moist savanna

Fig. 10.59 Vegetation in the eastern Himalayan altitude zones.

protected against radiant heat loss: some have leaves, others are hairy, like the Edelweiss.

In most systems, the vegetation varies from one high valley to the next, and vertical zones merge. The actual plant associations depend on steepness, soils, and micro-climates. Old and new screes bear different vegetation according to their origins, porosity, and age *in situ.*

PART 1 GLOSSARY

Terms related to specific topics are mostly defined where they first appear in the text. Here are some which are used more generally and may need some clarification:

adiabatic process: one occurring without gain or loss of heat from outside

adsorption: the attachment of an ion to a charged surface in such a way that it may be replaced (exchanged)

advection: the process which transfers a mass of warm or cold air horizontally

albedo: the proportion of the solar energy received by a particular surface that is reflected

autotroph: an organism which can manufacture its food from inorganic matter

biomass: the weight of organic matter per unit area (usually expressed as dry matter)

boreal: of the north

cation exchange capacity (CEC): the amount of adsorbed cations that can be retained (for exchange) by a given quantity of soil

centrifugal force: the outward force on a body rotating in a circle about a central point

climax vegetation: the final relatively stable stage of a successional sequence, established in response to natural conditions in a particular environment

colloid: a mixture of very fine liquid and solid particles whose great surface area has electrical attraction for ions

edaphic: related to the soil

eluvial: down-washing of colloids and fine particles in free-draining soil

ephemeral plant: one with a very short life cycle

epiphyte: a plant attached to other vegetation but obtaining nutrients from the atmosphere and decaying plant matter and not from its host plant

forb: a broad-leaved herb, as distinguished from a grass

friable: easily crumbled

genetic: relating to inherited characteristics of organisms

geostrophic flow: a resultant air flow produced by the interaction of force due to the pressure gradient and that due to the earth's rotation

geostrophic wind: one which in the free atmosphere responds only to the pressure gradient and the Coriolis force, causing it to blow parallel to the isobars

gneiss: a coarse-grained metamorphic rock with a banded crystalline structure

halophyte: a plant able to tolerate a relatively high environmental salt content

hygroscopic substance: one having a tendency to absorb water

illuvial: deposition in a soil of matter washed down from above

inversion: a situation where environmental temperature increases with altitude

ion: an atom or group of atoms having an electrical charge as a result of gaining one or more electrons (a negative **anion**) or losing them (a positive **cation**)

langley: a unit of solar radiation (one gm cal cm^{-2})

net primary production (NPP): the rate at which organic matter is accumulated in plants in a given ecosystem in a given time (dry weight per unit area)

organic matter: that forming the mass of living things, or derived from the decomposition of plants or animals

orographic: related to outstanding relief – hill slopes, mountains

plinthite: iron-rich concentrations in soil, hardening into a rock-like layer by repeated wetting and drying

pyrophyte: plant having adaptations which enable it to withstand fire (for many fire is necessary for regeneration)

rhizoids: very fine root hairs which may act to attach some rootless plants, such as mosses

sclerophyllous: with leathery evergreen leaves, protected from water-loss by a thick epidermis with a waxy cuticle

sere: a series of plant communities which change in composition as they progress towards a stable (climax) state of mature vegetation

sesquioxide: a compound with three oxygen atoms combined with two metallic ones

silcrete: a hard surface or near-surface layer cemented largely with silica

soil–water budget: a system evaluating the periodic balance and relative amounts of precipitation, evapo-transpiration, soil–water storage, water deficit, and water surplus at any place

sub-climax vegetation: a relatively stable stage of a successional sequence which has followed the clearance of the former vegetation

succulent: plant with tissues able to hold water within its thickened leaves or stem

synoptic weather map: with available detail plotted to give an overall view of weather phenomena over a given area at a particular time

trophic level: the designation of groups of organisms in an ecosystem according to their food sources (see autotroph)

vascular: with fine ducts (veins) which convey fluids

vorticity: a measure of the amount of spin of a rotating body; defined both in terms of its angular velocity and of the location of the vertical axis about which it rotates; and of whether it rotates in the same sense as the earth (positive), or not (negative)

xeromorphic: plant structure which enables it to withstand drought

Part Two

Disturbing the Ecosystem

11

SHIFTING THE BALANCE

11.1   Regulating the cycling of nutrients

The variety of life-forms in the major biomes, and the many different ways in which nutrients are exchanged between rocks, soils, and plants are related to complex controls over the processes of recycling. The productivity of the biome (NPP) often depends more on the **rate of cycling** than the amount involved.

The nature of soil nutrient exchanges and many of the processes involved are extremely intricate. Unfortunately, as the human population increases, and as we apply new forms of technology to the land, we unwittingly upset these processes, with knock-on effects through the ecosystem.

As we have seen, those plants at the lowest trophic level, the **autotrophs**, which can manufacture organic substances from inorganic compounds like carbon dioxide and water, also require minerals from the soil for cellular development. The absorption of these minerals is made easier by the process of **chelation**, which enables organic molecules from the products of decomposition to form complexes with metal ions – calcium, magnesium, iron, etc. (Fr. *chele* – claw, ie grasp). The complexes are more soluble, and usually less toxic, than other metal compounds; making it less likely that harmful amounts of an element will be absorbed.

Specialist bacteria also favour nutrient absorption. They have their own particular functions within the soil or on plant roots. As micro-organisms they act to regulate their own environment, and in doing so they convert compounds containing the elements needed for the development of cell protoplasm into forms which can be absorbed.

Plants require some nutrients in relatively large quantities: carbon, hydrogen, oxygen, nitrogen, potassium, calcium, magnesium, sulphur, and phosphorus are all **macro-nutrients**. But minute amounts of **trace elements** are also necessary for particular biological functions. If these are unavailable, the plants may wilt. On the other hand, if plants receive them in toxic quantities, or if they are radioactive, the plants are poisoned, or acquire lethal doses.

The roles of chelation processes and of bacteria are of immense importance in regulating the availability of inorganic elements, and to some extent in controlling the quantities needed for healthy development. In supplying a single element, several different types of bacteria may be involved, each with a specialised activity. Through ignorance, it is easy for us to destroy such vital micro-organisms. The bacteria themselves require aeration and are vulnerable to poisons. Unfortunately most people are unaware of how their actions may hamper natural cycling processes until the effects are visible and disturbing: when, for instance, excess fertiliser reaches streams and creates algae blooms which deprive other organisms of oxygen (p.147).

11.2   The cycling of macro-nutrients

Sulphur is an essential constituent of amino-acids in proteins. It is continuously cycled between life-forms and the soil by processes shown in Fig. 11.1 Micro-organisms acting on dead organic matter and excreta make some soluble sulphates immediately available. Hydrogen sulphide (H_2S) is also released, and under **aerobic conditions** (with sufficient oxygen) certain bacteria convert it to sulphur, which others then involve in processes producing sulphates, available for absorption by plants.

Under waterlogged, **anaerobic conditions**, unoxidised gaseous hydrogen sulphide forms iron sulphides. But this process tends to make another vital macro-nutrient, phosphorus, available to the plants, by converting insoluble ferric phosphate into a soluble compound.

Micro-organisms thus play key roles in the sulphur cycle and at the same time assist in cycling phosphorus. This further emphasises the need to avoid the unwitting destruction of minute, but essential life-forms.

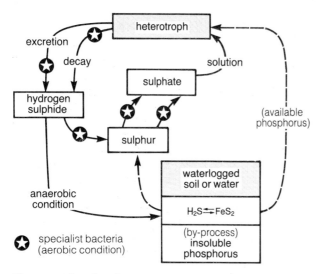

Fig. 11.1 The roles of various specialist bacteria in recycling sulphur between living organisms in the soil.

People themselves cycle sulphur in potentially harmful ways. Sulphur occurs in fossil fuels (of organic origin); so that we release vast quantities to the atmosphere as gaseous waste, with consequences discussed on p.147.

Phosphorus, as another element in the nucleic acids, is also essential for the majority of life-forms. Most organisms have mechanisms which regulate the phosphorus content of their protoplasm. They enable small quantities to accumulate to desirable amounts.

Phosphorus enters the soil from the weathering of phosphates in rocks, from volcanic fall-out, and from sea-spray. But it is not over-abundant in the earth's surface. Also, during its global cycling processes, it tends to accumulate in sediments in ocean depths. As soil erosion increases, and more and more particles are swept into the oceans, so

Fig. 11.2 Phosphorus cycling on a global scale. Much is retained for long periods in terrestrial rocks and marine deposits.

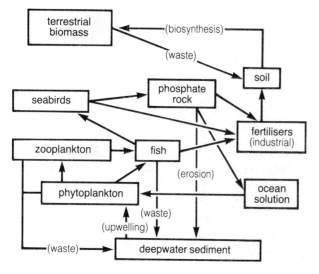

more phosphorus becomes unavailable, except in the very long term.

Rock phosphate is mined for use as fertiliser, but there is a limit to this resource. Therefore large amounts of fish are converted to phosphorus-rich fertiliser. Its place in the phosphorus cycle is shown in Fig. 11.2. The oceans, therefore, are much involved in this cycle. Where water upwells from great depths (p.46), it brings phosphorus to the phytoplankton zone, where it enters the biomass of fish. Most of their remains fall back to the ocean depths. But fish-eating birds excrete matter rich in phosphorus (guano); and where arid conditions allow such deposits to build up great thicknesses, as on the islands off Peru, they are removed as a valuable source of fertiliser, re-entering the cycle at a different place; but, of course, this needs controlling. Billions of anchovy are taken from the Peru Current for fertiliser (p.47). In other waters large-scale fishing supplies an abundance of phosphorus for human consumption.

Nitrogen is a gas contained in the great atmospheric reservoir; but continuously passes in and out of the air, in a cycle involving organic and inorganic compounds. Bacteria and blue-green algae can convert (fix) atmospheric nitrogen to soluble nitrates, which can be absorbed by plants. In legumes the **nitrifying bacteria** live in the roots. Nitrogen is also fixed to a lesser extent by lightning during thunderstorms; and, in a similar way, is artificially fixed and converted to nitrogen compounds by electrical-industrial processes.

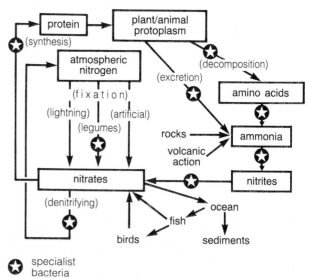

Fig. 11.3 The role of bacteria in the global cycling of nitrogen.

In the soil, other **denitrifying, anaerobic bacteria** transform nitrates and other nitrogen compounds into gaseous nitrogen. In the oceans nitrogen is cycled in aquatic ecosystems, and deep sediments retain a proportion.

Fig. 11.4 Trees throw shadows on the surface of algae choking the small canals, its growth stimulated by nitrogenous wastes.

We interfere with the nitrogen cycle by producing hundreds of millions of tonnes of nitrogenous fertilisers per year. We also release comparable amounts into the atmosphere through the partial combustion of fossil fuels, with the potentially disastrous consequences discussed below.

The addition of nitrogeneous fertilisers to agricultural land, the concentration of excreta from intensive cattle farming, and the rapid mineralisation following the clearance of natural vegetation all provide an excess of nitrates, which may wash in to streams and rivers, and accumulate in the relatively still waters of lakes. Such concentrations promote algal growth to the extent of depriving other life-forms of oxygen (by excessive **eutrophication**). High concentrations of fermentable organic matter also destroy plants at different trophic levels (by **dystrophication**).

11.3   Pollution of the local atmosphere

Sulphur **from fossil fuels**, of the order of 100 million tonnes per year, is released into the atmosphere in the form of sulphur dioxide, especially from power stations which burn coal and heavy fuel oils. In a few hours it combines with oxygen and water vapour to form sulphuric acid and sulphates, which are returned to the surface in forms of precipitation, mostly as **acid rain**.

Close monitoring of the pH value of the rainwater over western Europe, north-eastern USA, and Japan has shown values consistently below 5.6, compared with near neutrality in the mid 19th century. In both Europe and North America rain has shown pH values below 2.5, and in many cases the $SO_4 =$ ion has accounted for well over half the acidity.

Undoubtedly much of this is transferred into the atmosphere from factories in major industrial regions and from their power stations. The use of high chimneys to reduce local pollution has increased the concentration in general atmospheric circulation. Prevailing winds make it possible for pollutants to be carried from industrial regions in one country to create problems in others. Britain, perched on the western edge of Europe, is said to export two-thirds of its acid rain and to receive only an eighth as much. De-sulphurisation 'scrubbers', designed to remove a large proportion of the acidity at source, have been fitted to a number of power stations, but by no means all. Britain's output of acid rain from industry fell by a quarter from 1980–88; but, like other countries involved in expensive and lengthy control operations, it continues to inject sulphurous emissions into the atmosphere.

All plant life is adversely affected when a particular threshold of $SO_4=$ concentration is reached, and **conifers and lichens are very sensitive**. In Scandinavia especially, the pines have suffered badly. Bare lunar-like landscapes in

Fig. 11.5 Smoke pours from small Beijing factories, hugging the surface under the mid-winter inversion conditions. This is a time when outblowing winds bring dust from the interior. At present the city has no major problems from vehicle fumes.

Fig. 11.6   A thick layer-cloud lies over the Blue Mountains in New South Wales; beneath, at Port Kembla, smoke from factories and metal works rises while still hot and then sinks, drifting sulphurous fumes along the valley and industrial districts beyond the ridge.

western Tasmania emphasise the extent to which sulphur dioxide emissions can directly affect an area of dense natural temperate forest, in this case from copper smelting.

The increasing acidity of lakes over such rocks as granites and quartzites, which are naturally acidic, damages the ecological balance. Fishing is affected; and many small lakes in Canada and Scandinavia have become almost devoid of plant and animal life. In Scotland, sharp bursts of acidity from snow-melt and rain storms are seen as the main contributors to the death of salmon in the headwaters of the rivers Spey and Dee.

Sulphurous fumes also endanger the respiratory system of animals, including humans, as do many of the nitrogen compounds which enter the atmosphere, mainly from **the partial combustion of fuels in vehicles**. The oxides of nitrogen, peroxyacetyl (PAN), and benzopyrene are all either emitted or created in the local atmosphere; they are harmful to lungs, increase the risk of cancer, and damage plant tissues.

With strong solar radiation ozone production allows the combustion products to combine with other pollutants and form **photochemical smog**, containing dangerous secondary pollutants, such as PAN. Los Angeles, a city with a high vehicle density, and in summer a high intensity of short-wave radiation, combined with subsiding air, is notoriously affected by such smog; but many other cities also experience it, from Mexico City to Athens.

11.4   The local distribution of air pollutants

A plume containing pollutants emitted from a single high stack, at a relatively high velocity, and at a temperature above that of the environmental air, makes for an initially bouyant emission, which prevents pollutants reaching the ground close to the source. However, wind strength and

the stability or instability of the local atmosphere affect **the plume pattern**, as shown in Fig. 11.7:

A   Local instability, as during a hot summer afternoon, creates movements which carry the pollutants up and down in loops of increasing size; and may pollute the ground relatively close to the source.
B   Under fairly windy, but stable, conditions the vertical and lateral spread are about equal, so the pollutants fan out into a cone-shaped plume.
C   When the lower air begins to become stable, as the surface cools in the early evening, for example, the plume remains elevated and tends to be dispersed upwards.
D   Under inversion conditions in the upper air, as the ground begins to heat up the gentle looping movements of the plume have an upper ceiling, and so bring pollutants to the ground at intervals along the line of flow.

Fig. 11.7   The distribution and dispersal of smoke particles under various meteorological conditions.

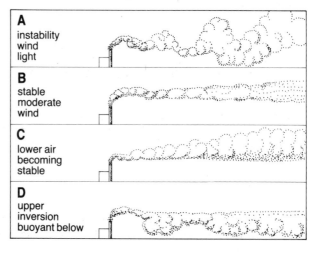

A
instability
wind
light

B
stable
moderate
wind

C
lower air
becoming
stable

D
upper
inversion
buoyant below

Over long distances, perhaps upwards of 10 km, the individual plumes lose their identity and simply add to the more general atmospheric contamination. **Urban areas**, with numerous individual places of emission, act as a large single source, whose plume may extend, perhaps, for 100 km or more downwind. While **on a continental scale** pollutants from individual industrial–urban areas may increase the general load carried by a contaminated airstream, and perhaps be transported over thousands of kilometres.

11.5   Pollutants and global circulations

A number of global cycles may be so disturbed by human actions that the impact affects conditions far from the event, and perhaps far into the future. This is most apparent in **interference with the hydrological cycle**, shown in Fig. 3.2, where about nine-tenths of the surface water draining to the sea is now a vehicle for the transport of wastes from human activities.

The water in river systems is used to dilute sewage and industrial effluents. Even a large river like the Rhine transports chemicals, including heavy metal compounds, and becomes an industrial sewer capable of corroding the vessels which ply it (and add their own pollutants). The shallow North Sea is highly contaminated, not only by contributions from the Rhine, but by discharges from all the highly industrialised countries which drain into it, and by direct dumping of wastes from ships. Similar contamination affects the Great Lakes in North America, where there are already problems of eutrophication.

Atmospheric contamination and interference with the balance of the constituents of the atmosphere are less easy to recognise or assess quantitatively on a global scale. But our **interference with the carbon dioxide cycle**, through the emissions from fossil fuels and the effects of clearing large areas of carbon retained in natural vegetation, is on a sufficient scale to require careful monitoring and possible controls.

The burning of fossil fuels, in particular, and clearing and burning forest have immensely increased the emission of carbon dioxide into the atmosphere. The table below stresses the results of increasing fuel consumption since World War II.

Table 11.1   Carbon emission (million tonnes)

region	1950	1980
North America	750	1300
E. Europe/USSR	300	1250
Western Europe	420	900

In North America and Western Europe carbon emission has recently levelled off, or fallen. But it

continues to rise elsewhere. The developing countries are rapidly increasing their total emissions into the atmosphere as their fuel consumption responds to urbanisation and industrialisation. Their overall emissions rose some twenty-fold between 1950 and 1980.

Increases in the concentration of atmospheric carbon dioxide during the last century are well documented. But there is continual recycling between the many carbon reservoirs (Fig. 11.8), so that the *extent* to which one can attribute the increase to human activities is difficult to assess. Nevertheless the carbon dioxide released annually to the atmosphere from fossil fuels has increased by about an eighth since the start of the century.

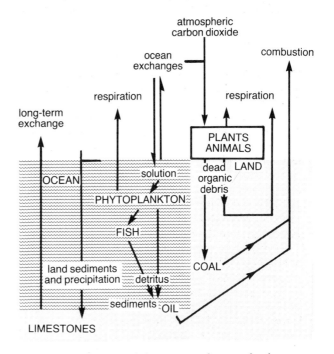

Fig. 11.8   Long-term and short-term exchanges of carbon dioxide and its global cycling.

The prime concern is that the increase in carbon dioxide atmospheric content will enhance the **greenhouse effect** (p.8), and so considerably raise the temperature at the earth's surface. As there is also an input of heat from our cities, industries, and vehicles, it has been suggested that the mean terrestrial temperatures might rise by 10 C°; which would mean the total melting of the polar ice caps and a rise in mean sea-level of some 80 m, with the consequent flooding of urban concentrations on the coastal plains. This is at the scare-level of prediction. A number of more realistic predictions have seen a likely rise of the order of 1 C° over the next half-century, with the high latitudes experiencing a rather greater increase. However, the former prediction may act to make people aware that there is a problem.

Fig. 11.8 shows exchanges continuously taking place through **cycling between the world's major carbon reservoirs**. They act on different time-scales. Some responses are rapid, such as respiration releasing carbon dioxide to the atmosphere; others, such as the transfer of carbon dioxide from the atmosphere to the ocean, are very slow. But each of these exchanges contributes towards a dynamic equilibrium.

In trying to estimate the effects of deforestation, agricultural developments, and fossil fuel contributions, we have to consider that the world's climate undergoes long-term and short-term changes. This complicates predictions. For example, the exchanges between atmosphere and ocean depend on the nature of the sea-surface temperatures and oceanic circulations at a particular time. Although the oceans have a capacity to absorb extra carbon dioxide it is likely to be of the order of several thousand years before the quantity estimated as due to human activities over the last century would be absorbed. On a century time-scale about a third would go into the ocean and about a tenth into vegetation. There are also feed-back factors: for instance, carbon dioxide is less soluble in warm water than in cold, so that surface-warming acts against the absorption effect.

In attempts to predict the future, **climate models and analogue models** (comparing past occurrences) have been used: when averaged, they suggest that a doubling of the carbon dioxide atmospheric content would increase global temperatures by 2.5 C°. This has led to predictions that millions of agriculturalists would suffer, especially those farming marginal areas in the tropics and sub-tropics; and that droughts would jeopardise grain production from the temperate grasslands. Yet hundreds of sound agronomic experiments have established that carbon dioxide enrichment acts as a stimulus to plant growth and development. It also induces partial stomatal closure, so plants lose less water by transpiration. Agronomists and water authorities have recently separately concluded that by doubling the carbon dioxide atmospheric concentration **crop yields would probably increase** by a third (though not necessarily the nutritive value), and that the yield per unit of water used would probably double: conclusions of great significance for the less developed countries.[3]

At the moment it has been established that the amounts of carbon dioxide in the atmosphere have been steadily increasing, but not that there have been related temperature increases; for annual, decadal, and longer-term variations are superimposed over each other. If an increase *is* detected in the next decade, it may still have been due to natural causes; and the debate as to whether the effects will prove catastrophic overall, or even beneficial in some respects, will continue. In the meantime it seems sensible to try to control the inputs of carbon dioxide as far as we are able. They have been proportionally slowing in the developed countries, but increasing in the less developed.

The chlorofluorocarbons (CFCs) used in refrigeration and aerosol-can propellants are finding their way into the stratosphere, and not only adding to the greenhouse effect, but are **reacting destructively with ozone**. Other halogen-carbon compounds widely used as solvents are also increasing in the upper atmosphere. There is a cycle of destruction of ozone (O_3) by chlorine, with the help of ultraviolet radiation (UV).

$$Cl + O_3 \rightarrow ClO + O_2$$
$$O_3 + UV \rightarrow O_2 + O$$
$$ClO + O \rightarrow Cl + O_2$$

so that $2O_3 \rightarrow 3O_2$ takes place, and the ozone has been lost to the atmosphere.

This is a potentially serious occurrence, for ozone intercepts much of the incoming UV, and its depletion leaves life-forms more vulnerable to additional strengths of radiation. The effects on plant tissues would affect crops, and in humans there seems to have been a recent increase in skin cancer, which some attribute to a thinning of the ozone layer.

Apart from the general effects in the stratosphere, since 1979 the concentration of ozone high above Antarctica appears to have fallen by some 40 per cent, with an expanding hole in the ozone layer over the South Pole. The difficulty in attributing this to CFCs, or to any other single cause is that scientific observations based in Antarctica extend only over a short period. Periodic upwellings of the lower air may be connected with this effect; as may a burst of intense solar radiation, such as occurs from time to time. Investigations are also taking place into the possibility of similar occurrences over the Arctic polar regions.

A biochemical influence also stresses the complexities of feedbacks in the global system. More than half the chlorine entering the atmosphere, and about a quarter of that reaching the stratosphere, is derived from **naturally produced chloromethane**. Large quantities of this are biosynthesised by wood-rotting fungi. Scientists have pointed to the fact that large-scale clearance of rainforests, undesirable from many points of view, results in the destruction of these fungi, and has already reduced the atmospheric concentration of chloromethane: so that the arrival of the CFCs is, in a sense, compensation. Nitrogen oxides from denitrifying bacteria are also involved in the dynamic equilibrium of stratospheric ozone. Nevertheless, agreements to restrict the use of CFCs are sensible; for although the ozone layer would very quickly be reformed in the absence of destructive agents, the concentrations of the latter have been increasing.

11.5 Cycles of climatic change

Climatic changes have been occurring throughout earth's history, with irregular fluctuations between periods when the mean global temperature was markedly higher or lower than before. Long-term and short-term changes continue, and in trying to estimate the effects of human interference with the atmosphere, we should at least appreciate the periodicity and likely extent of such natural fluctuations.

Geological observations give evidence of glacial events thousands of millions of years ago. **Modern techniques** involve a comparison of the proportions of oxygen isotopes in water and in contemporary fossil carbonates, as revealed in core samples from ice sheets and ocean beds. They allow us to follow temperature changes in more recent epochs. Pollen analysis, sedimentary core samples, and tree ring comparisons are backed up by increasingly sophisticated methods of acquiring climatic data.

The late Mesozoic period, some 70 million years ago, was so warm that there were no polar ice caps. But during the Tertiary period the mean temperature decreased almost continuously. Ice first formed around Antarctic shores about 38 million years ago; and, after a warmer period, ice sheets began to cover Antarctica some 14 million years ago. About 12 million years later the northern ice sheets developed.

Fig. 11.9A shows the alternating glacial and warmer inter-glacial periods of the latter part of **the Pleistocene**, representing climatic fluctuations which have occurred over **the last 2 million years**. It also shows that **we now live in an unusually warm period; only for a small fraction of this time has the earth been as warm as it is today**.

If we make our time-scale **the last thousand years**, another remarkable sequence of climatic fluctuations appears (Fig. 11.9B); but notice that the amplitude of swing is only about 1.5 C°, compared with 5–6 C° over the period shown in Fig. 11.9A. Yet the effect has been to make people say that Europe experienced a little ice age, setting in during the 14th and 15th centuries. Fig. 11.9C

Fig. 11.9 Variation in mean air temperatures in the northern hemisphere, estimated by various methods for different periods of time. Notice the ranges between the maximum and minimum estimates in each case.[6]

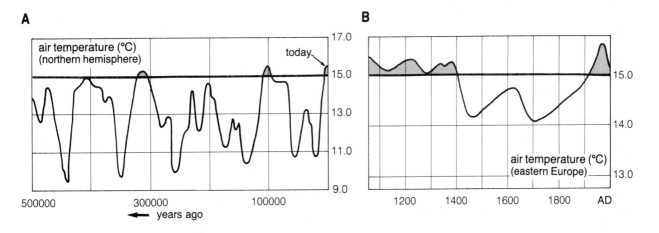

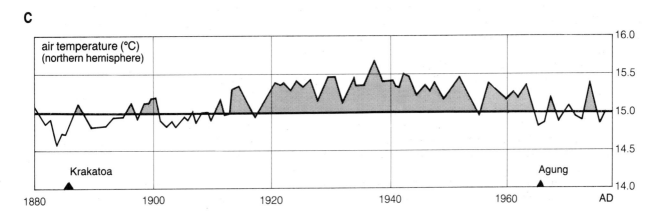

shows how temperatures have continued to fluctuate during the last century. The generally cooler conditions of the end of the last century were replaced by a warmer period up to the middle of the present one. Similarly, unusually wet and unusually dry periods have alternated, affecting geomorphological rates of change; and, as shown in Fig. 11.13, can be potentially disastrous where a growing population puts pressure on the land.

Long-term cyclic variations in climate, of the order of tens and hundreds of thousands of years, appear to be related to long-term changes in the intensity of solar radiation reaching the upper atmosphere. This is linked with changes in the earth's spatial position, the eccentricity of its orbit, and its tilt relative to the plane of the ecliptic. **Fluctuations in solar activity** also seem to correspond to glacial and inter-glacial periods. While in the short-term the absence of sunspots seems to be associated with cooler spells, usually with colder winters.

We have seen that water vapour, carbon dioxide, aerosols, and dust are capable of absorbing much long-wave radiation, so that their atmospheric concentrations are significant short-term climatic controls; as are events which modify the earth's albedo (p.8).

Volcanic activity can project an excess of dust into the upper atmosphere. Fig. 11.9 shows that when, in 1883, the explosion of Krakatoa, in Indonesia, ejected vast quantities of dust into the atmosphere, it appeared to cause a fall in the earth's temperature of the order of 0.5 C°. Direct observations have shown that the eruption of Agung, in Bali, in 1964, both lowered the temperature of the troposphere and raised that of the stratosphere, where minute particles increased the absorption of incoming radiation.

As always, **feedbacks in the global system** are complex and far-reaching in effect. For instance, both dust and water in the atmosphere may add to the greenhouse heating effect, with increased drying and aridity in some areas. But dust particles are also nuclei for the condensation of water vapour, so increasing the cloud cover, possibly with precipitation and local cooling at ground level.

On the long-term scale, glacial periods, which lock up so much water of the hydrological system as ice, cause a worldwide fall in sea-level. Land-masses link and allow fauna to migrate, especially away from the increasing cold of the higher latitudes. In such cooler periods, with higher precipitation and lower evaporation rates, lakes appear to have a greater volume. As the position of the climatic belts shift, so do the vegetation zones. The expansion of deserts during inter-glacial periods is clearly indicated by fossil dunes in places which today are covered with grassland, shrubs, and low trees, as in the Kalahari.

11.6 Climatic fluctuation, population pressure, and desertification

It is in the marginal lands bordering the truly arid areas where climatic fluctuations combine with overpopulation by humans and livestock to create the twin disasters of starvation and the loss of land. The result may be an extension of the desert, which can prove irreversible – **desertification** – though there are remarkable instances of regeneration of vegetation and recovery of impoverished land. These remain areas of cycles of disasters and recovery.

This has been the fate, over the last two decades, of millions of families who live in **countries of the Sahel belt**, stretching across the southern parts of the Sahara, from the Atlantic to the Indian Ocean; and of those farming the high, dissected lands of the Ethiopian plateau.

There can be many consecutive years when the rainfall is above average (Fig. 11.12) and the nomads and semi-nomads can move their tens of millions of cattle, sheep and goats over the dry grasslands, where bushes and low trees provide nourishment for browsing animals. Daily they congregate at wells or water-holes, which are a few kilometres apart. There are villages where branches of an inland delta, or old Pleistocene flow channels, retain sufficient water from the summer flush to support subsistence agriculture.

Reliable rains during the 1950s saw agriculture and pastoralism both extend northwards. The natural rate of population growth is high; and farmers and nomads alike greatly increased their numbers of livestock. But even in good years, the dangers from overstocking is apparent. A new borehole can encourage them to keep even more animals, irrespective of the carrying capacity of the vegetation. Trampling about the watering places causes soil erosion; and there are many examples of the complete removal of vegetation radiating out from a water source. Also, with an

Fig. 11.10 The moist south-westerlies bring periods of summer rain to the Sahel and the Ethiopian highlands. But for years at a time there is insufficient precipitation for pastures or crops.

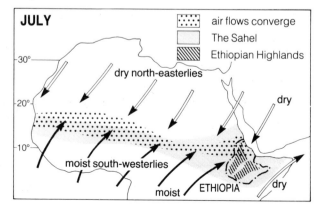

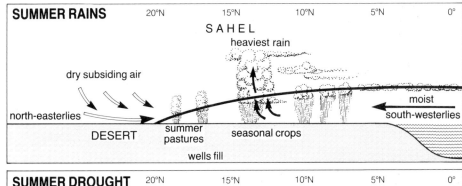

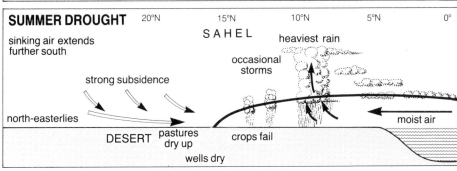

Fig. 11.11 Sections which emphasise again the circumstances which make it impossible for the Sahel to receive adequate rainfall. Such conditions may persist for years at a time. (See also Fig. 4.29.)

increasing population, people tend to remove more woody tissues for fires.

Fig. 11.11 shows, in a simple way, how the deterioration of pastures and the failure of crops are linked with the northward limits of the summer rains.

As drought sets in, the animals quickly exhaust the normal grazing and turn to plant species they usually ignore. As one drought year follows another, nomadic families begin to migrate southward to more settled areas. But the population of these marginal farmlands has also increased; and recent droughts have extended even further south. Today, many migratory routes are blocked, particularly at international frontiers. There has been severe starvation within the Sahel, and millions of animals have perished.

There is **a downward spiral towards desertification**. Soils which are overused, deteriorate, and are slow to recover when climatic conditions improve. Their carrying capacity is lower. The spiral continues: for though climatic vagaries may

be a principal cause, progressive desertification can itself affect the climate. The bare land increases the albedo – reflects more of the incoming radiation. This may enhance the sinking effect of the air, increasing the aridity and lessening the chance of convective cloud formation.

Any climatic event can be seen as the result of previous occurrences and the cause of later ones. On a global basis, droughts and desertification may be responses to far distant events (p.46). The north–south movements of the ITCZ and the changing path of the upper easterlies may have links with the Indian monsoon, and even with the Walker circulations; but the links are far from clear. There must be connections also with the causes of the long- and short-term global temperature fluctuations; and, to a lesser extent, with the self-perpetuating effects of desertification. There are also, of course, the human factors: in Ethiopia, and elsewhere, political instability hinders long-term projects to check, and then prevent, the increasing desertification.

SAHEL - VARIATIONS IN ANNUAL RAINFALL standard deviation (selected places)

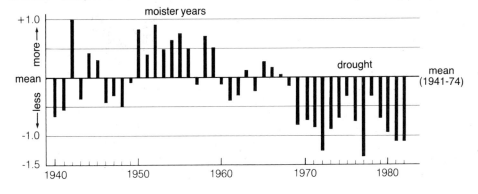

Fig. 11.12 Periods with above average rainfall, which encourage over-stocking, and with below average rainfall, which then make for desertification (after L. Musk[4]).

Fig. 11.13 Here, as in most of the Sahel, a succession of good years encourages settlement. Communities like these in the Ethiopian highland live on the brink of disaster. After years of plenty, the long drought has turned many of the fields to dust. The water-table is low, and stream beds and many wells are dry.

11.7 Responses to rainforest destruction

The immediate dangers in clearing tropical rainforest are already apparent from the description of forest nutrient cycling on p.114.

(a) The same chemical nutrients are continuously and rapidly recycled, with fungi as the main agents.

(b) The upper soil humus layer is slight.

(c) The root system is shallow, but acts as an efficient filter for dissolved substances.

(d) The tree canopy physically protects the surface.

(e) The canopy water and stem run-off are rich in nutrients, many derived from fauna, which are an essential part of the recycling process.

(f) Soil particles are largely of quartz and kaolinite, mostly unable to retain nutrients.

(g) The number of identified individual plant and animal species runs into millions.

(h) Plant and animal species interact to form ecological niches within the forest.

(i) Most of the extensive rainforests have developed over millions of years – though over large parts of south-east Asia there is secondary forest.

Locally the effects of clearance are instant nutrient loss, exposure to the impact of heavy rain, the selective removal of small particles, and the erosion of the sandy remainder. Rivers receive a much greater sediment load, causing bed silting and flooding. **Impoverished soils prevent forest regeneration**.

Globally, large-scale deforestation has many undesirable aspects. The destruction of an ecosystem with so many interactions and feedbacks among the life-forms means the loss of a uniquely rich biome. More than this, **it depletes global genetic stock**; and, as new species are still being identified, the losses are incalculable.

The disruption of local populations with a life-style adapted to their environment has gone so far that it is difficult to find satisfactory solutions. For example in the exploitation of Amazonia for agricultural–commercial–industrial pur-

Fig. 11.14 A sloth in the Brazilian rainforest protects itself by its motionless behaviour, and by the green colour created by algae which, in the hot, damp conditions, thrive in the grooves of its coarse hair. The algae are food for the larvae of moths which lay eggs in the hair. Here is a symbiotic community, illustrating that the destruction of some life-forms endangers others; and in this case the sloth itself depends on leaves from the *Cecopria* trees.

Fig. 11.15 Severe gulleying two years after rainforest clearance in south-west Antigua.

Fig. 11.16 The same site six years after; the gulleys are deeper and wider; few plants root in a soil deprived of nutrients.

poses by non-indigenous peoples, the newcomers have opened new, usually false, horizons for the growing indigenous population. It is difficult to avoid frustrating the ambitions of the rainforest inhabitants, or, on the other hand, creating a 'zoo' situation if they are confined to separate designated areas of unexploited forest.

As deforestation continues, **the release of carbon** from the forest biomass must be viewed in terms of complete clearance. Amazonia's forests contain about 20 per cent as much carbon as is present in the entire atmospheric carbon dioxide. Replacement by a crop-grass form of vegetation would release about 80 per cent of this to the atmosphere.[5] The implications have already been discussed.

Deforestation also means that **evapotranspiration losses decrease** and reduce the total annual rainfall. It also makes for greater seasonal variation in humidity, so trees in remaining forest areas might be deprived of groundwater for longer periods, and may not survive.

These are *optimum* situations, based on complete forest removal in Amazonia. But, of course, deforestation proceeds in other parts of the world, and the current effects of lumbering in Borneo, for example, are just as disturbing.

Nevertheless, with care, and using long-developed agricultural systems and techniques, **many parts of the tropics have been converted to productive agricultural areas**; notably the terrace systems of south-east Asia and Indonesia, with their careful water control and methods of nutrient replacement (Fig. 5.6). Exchanges of water with the atmosphere from the large areas of terraced crops, and from irrigated surfaces, are still considerable. **Stable ecological systems have evolved**; but even these are under pressure from increasing human populations and the introduction of chemical fertilisers and insecticides.

11.8 Injecting chemicals into the system

Many of the pollutants entering the global atmosphere circulation become so diluted that they are regarded chiefly as local irritants. Some, however, are **cumulative poisons** such as the lead compounds released from petrol additives. They are acquired in small amounts, but may gradually build up toxic concentrations in body tissue.

The increasing use of chemical fertilisers, pesticides, and defoliants has fed a number of potentially dangerous substances into food-chains. Chlorinated hydrocarbons used as pesticides, of which DDT has been the most widely distributed, have spread harmful effects to plant and animal life far from the point of application. They may move into water in very small amounts, but eventually enter the human system in much higher concentrations.

Selected water plants were found to absorb pesticide and acquire a cellular concentration of 0.1 parts per million; but fish feeding on them built up a concentration in their tissues of 1.1 ppm; while in fish-eating birds the concentration rose to 4.0 ppm. There are numerous examples of

Fig. 11.17 Intensive crop-spraying with chemicals on hillsides draining directly to the Rhine below. Chemical pollution takes place in one form or another along hundreds of kilometres of the main river and its tributaries.

Fig. 11.18   Heavy metal works and chemical factories line the transport routeway of the Rhine, and the continuous passage of powered vessels adds to the water pollution.

predatory birds succumbing to DDT poisoning; and sampling has discovered mean concentrations of 12–26 ppm in human fat tissues.

The consequences depend on the chemical acquired in such amounts, and need not be elaborated: sufficient to point to the process of **biological magnification** through the food chains within any biome. The dangers are emphasised by the fact that the toxic effects of sea-borne lead have already shown up in phytoplankton; which, at the base of many branching food chains, are the most numerous producers in the world at the lowest trophic level.

11.9   Local climate modifications

Apart from **siting crops** to take advantage of natural conditions, or avoid disadvantages – soft fruit grown on sunny slopes, rather than on the valley floor – some **farmers modify the local climate** to suit their activities. Fruit trees are protected against radiation frost by burners, which provide heat both in the typical inversion layer close to the ground, and an upper smoke surface to minimise radiation losses. Sometimes elevated fans are used to mix the cold lower air with warmer air above.

In some arid lands, where local areas need additional rain, clouds may be induced to release precipitation. Tiny particles of silver iodide or solid carbon dioxide are dropped onto clouds to act as freezing nuclei. Ice crystals form within the cloud and melt before falling to earth.

Greenhouses and cloches, for protection and heat trapping, in a sense **create a new local environment for plants**, rather than modify the local climate. Chapter 12 considers the climate in buildings and about groups of buildings, but in this case in an urban environment.

12

URBAN CLIMATES

12.1 Climate and human comfort

We need to maintain a body temperature of 37°C, and are comfortable with an environmental temperature of about 20°–25°C. We become particularly uncomfortable at low temperatures, and insulate ourselves from cold air by wearing additional clothes, or sheltering in an artificially heated room. We are able to carry on with our activities in a hot climate mainly because of the control of body temperature through sweating; though the body may not be able to lose water through the skin when the air is already saturated with water vapour.

I

air temperature °C	apparent temperature °C		
35	34	42	58
30	29	32	38
25	23	25	27
RH %	20	50	80

Table I shows **what the body senses the temperature to be** with various degrees of moisture in the air (RH = relative humidity). A combination of heat and humidity makes for heat exhaustion and discomfort ... a different condition to that caused by dehydration in a hot dry climate, when it is necessary to drink sufficient water and replace salt lost through sweating.

II

air temperature °C	apparent temperature °C		
+10	+9	0	−3
+4	+3	−8	−12
−12	−14	−32	−38
wind speed m/sec	2.0	9.0	18.0

Wind causes rapid evaporation of sweat and consequent cooling. When the air is cool, its movements sometimes cause **wind-chill**; a combination which makes for considerable discomfort. Table II shows the chilling effect of wind at various speeds, given as an **apparent temperature**, which is how people feel it to be. As in Table I, it is calculated from a formula based on the responses of a large number of people.

12.2 Buildings and intentionally modified climates

The first response to such conditions is the type of clothing adopted – a string vest designed to leave an underlying, insulating heat layer and check the loss of body heat to bitterly cold air; or a broad-brimmed hat to protect against the direct impact of solar radiation.

Buildings constructed to provide shelter from adverse weather elements and a comfortable atmospheric environment, also vary according to the nature of the elements. Here, too, insulating layers are extremely important. About the building itself a thin laminar air layer tends to remain in immediate contact with the surface – **a boundary layer, a few millimetres deep**. Being non-turbulent, this tends to diminish direct heat exchanges between the surface and the more turbulent air beyond, especially in still conditions. A wind force decreases the thickness of the layer and also creates turbulence about the buildings, and so diminishes the insulating value.

Consider **the nature of the energy exchanges** between the shell of the building, the interior, and the environmental atmosphere.

(a) At any time some walls and parts of the roof may receive direct insolation while others are in the shade. The illumination and the intensity will vary during the day.

(b) The roofing materials and aspect of the roof

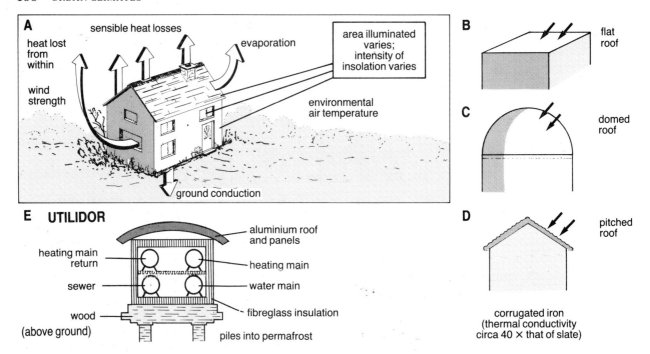

Fig. 12.1 Climatic influences on buildings.

are important, especially in the tropics – because of the high elevation of the sun (Fig. 12.1B, C and D).

(c) The albedo of building and roofing materials affects the absorption of energy by day and their thermal conductivity affects the radiation losses at night. Thus a corrugated iron roof may become hot by day, but also allows rapid cooling at night.

(d) The area and location of windows affect the penetration of heat to the interior.

(e) The interior gains heat from deliberate space-heating, from such activities as cooking, and from people and animals.

(f) Heat losses depend on the surrounding air temperature. The thickness of the laminar layer is important. The losses may be controlled by such precautions as having the roof well insulated from the interior ceilings; double glazing for windows; carpeting; good insulating building materials; wall spaces; and general air-tightness.

(g) Heat loss can occur as wet surfaces, or attached vegetation, dry by evaporation.

(h) Loss by conduction through the ground is usually fairly small, but can be high in very cold climates. In these areas buildings may be raised on stilts to minimise conduction losses. **In permafrost regions**, the provision of facilities like water, sewerage, and heat from a communal source, usually involves encasing the pipes in **an above-ground utilidor** (Fig. 12.1E).

Heat conservation and efficient air conditioning are important economic considerations for both householders and manufacturers, whose plant may be working round the clock. As Plate 13 shows, manufacturers may now use colour-coded thermal images obtained by airborne thematic mapping to pinpoint areas with a particularly high heat loss.

Of course, buildings also have to be designed to cope with the excessive heat and humidity in the moist tropics, and are equipped with such adaptations as shady verandahs, fan ventilation, latticed structures, space beneath floors to allow air circulation, and so on.

Climatic factors often influence **the exact siting of housing**. There are considerations of shelter, drainage and water supply. In Australia there is a very real threat of bush fires, which may extend into the suburbs of the large cities; therefore, where there is a choice of siting, lee slopes are favoured, for fires increase in speed as the heat stimulates strong upslope draughts. There, static water is located, if possible, above house level, to provide pressure in fire emergencies.

In south-east Australia the time of maximum danger is when hot, dry northerly winds set in ahead of an eastward-moving trough to the south. Bush fires then move southwards; but as the trough moves away, south-westerly winds set in, and so extend the fire-front sideways over large areas. A single spark can ignite vapour from the eucalypts, so that fire leaps suddenly from one vegetation clump to another.

Householders in Australia are instructed to maintain an area clear of undergrowth, twigs, and

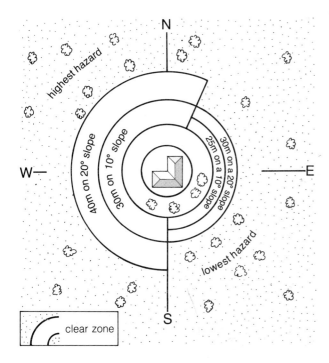

Fig. 12.2   In south-east Australia the area cleared as a precaution against bush fires varies with the wind direction and slope.

leaves for distances from the house which vary with slope and direction (Fig. 12.2), and to set sprinklers 10 metres apart around the house for spray overlap.

12.3   Urban areas and climatic modifications

A large urban area is more than a collection of buildings. It contains the human population and the vehicles and machinery which use, generate, and release energy and which discharge wastes. There are many different surfaces, of roads, paved areas, trees, parks, etc. The relative location of the buildings themselves create deep urban canyons between, and the maximum surface areas may be facing horizontally rather than vertically. The high-rise centres of the inner cities contrast with the suburban areas and the countryside beyond, and within them are innumerable micro-climates.

The urban area, as a whole, changes the general atmospheric properties of the region in which it is located; so that **urban climates differ from those of their rural surroundings**. Within the cities micro-climatic variations, such as the creation of locally strong air turbulence, have considerable impact on people, and even on structures. But first we shall look at the ways in which urban climatic conditions are at variance with those about them; though the climatic elements, of course, vary from city to city, with latitude, location, and season. Obviously urban–rural contrasts will be different for those areas with a snowy environment and those in the humid tropics.

As wind blowing across the countryside encounters the edge of a large urban area, it begins to pass over uneven clusters of buildings intermingled with open spaces – a surface which is usually rougher than that of the rural areas. Obstacles change the wind direction and speed, and the frictional drag causes turbulence. This extends through the overlying layer of air to a height where the regional wind continues to blow unchecked.

Fig. 12.3 shows the wind speed at the top of this **urban boundary layer** and the way the wind speed decreases until relative calm prevails in the shelter of the city buildings. In fact, this is misleading, for though there are calm, sheltered areas within the city, there are also places where buildings channel the air into high velocity winds. However, the overall effect may be seen as a decrease of wind speed with altitude through the urban boundary layer, and a comparison is made with that over the more open countryside.

The city's atmosphere below the roof-level 'canopy' can be seen as an **urban canopy layer**, which, in fact, includes **a complex intermingling of micro-climates**.

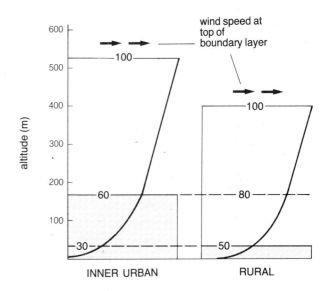

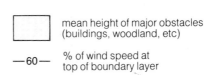

Fig. 12.3   A comparison of wind speeds in the rural and urban boundary layers, as the air is disturbed by surface features. There are many micro-variations within the urban canopy layer.

Fig. 12.4 shows, schematically, the creation of **an urban plume** extending from the windward edge of the city. In this the turbulence and convective movements tend to create cloud and increase precipitation downwind.

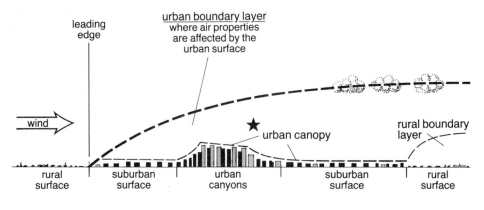

Fig. 12.4 A large urban area disturbs the overlying atmosphere, and creates a plume of air, whose properties are modified by turbulence, convection, pollutants, and local sources of water vapour. Clouds tend to form and precipitation increase downwind of the main urban area.

Water, cloud, and precipitation

The water storage in a city is generally less than in the rural areas, for run-off is greater, and gutters and sewers rapidly dispose of it. Water vapour is released into air over the city by fossil fuel combustion, and from cooling towers, open storages, and watered gardens; and water is imported to the city to supply residential and industrial needs. As there is less vegetation, there is less evapo-transpiration than from rural surfaces.

Urban air, polluted by small particles, favours condensation, and as a result droplets form in updraughts in the urban boundary layer. As it takes time for them to reach sufficient size to fall as rain, the areas downwind of the city usually show the greatest increases in precipitation.

The urban heat island

Air pollution in the urban atmosphere tends to absorb some of the incoming solar radiation. But in the middle latitudes the urban albedo is generally above that of the rural areas, and the surfaces absorb much heat. Also the dust particles and increased cloud-cover tend to close the atmospheric window (Fig. 1.3), and absorb and return long-wave radiation from the city beneath. Apart from heat stored in the materials, buildings are also heated by people and their energy-emitting devices (**anthropogenic heat**). However, long-wave radiation is not simply emitted vertically, but transmitted from building to building across narrow streets, with absorption losses in the process.

Compared with the countryside, there is less heat lost through evapo-transpiration, and a

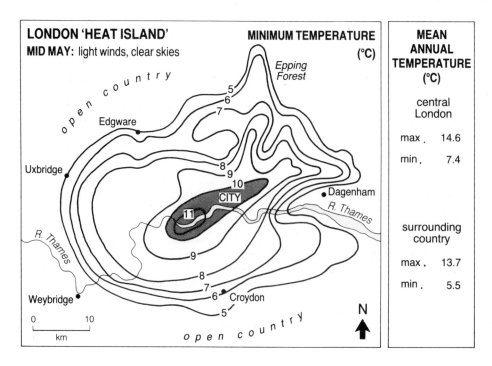

Fig. 12.5 The minimum temperatures on a cloudless day with little air movement point to a well-established heat island over Greater London (after Chandler[7]).

smaller loss of sensible heat due to the overall reduction in wind speed at lower levels. Taken together, all these causes make for the creation of a 'heat island' in the urban canopy layer.

The intensity of this heat island shows diurnal variations, usually being at a maximum a few hours after sunset and a minimum in the middle of the day. After sunset the urban area holds heat longer than the rural areas, with their direct loss of long-wave radiation and moist surfaces. The contrast between the two is greatest when the sky is open and the air still.

Fig. 12.5 shows **a heat island over London** under calm conditions, and mean figures which point to significant differences overall; though on a mild, windy day any differences are usually slight. Here the minimum temperatures show a typically steep rural–urban gradient, with a central tableland of warm air about the warmest central part of the city. There will obviously be variations throughout caused by such features as parks, water surfaces, denser and less dense housing, and industrial concentrations.

Topography and human responses affect the pattern, so that warmer conditions are seen to extend northwards along the Lea valley, with its housing developments, and a cooler wedge can be seen to the north-west, where there is more open country.

Eastwards, along the Thames, cool air over low-lying estuarine land shows up between the warmer air over dense urban districts to the north and south. Here shallow, thermally-induced winds are apt to move into these built-up areas on otherwise still days, quickly losing speed among the houses.

The regional wind speed and direction affect the intensity and pattern of the heat island. When light westerlies blow along the line of the Thames valley the higher temperatures are displaced eastward, producing steep marginal gradients to the east.

The characteristics of the heat island may change with the years; when, for instance, the nature of **urban-industrial air pollutants** changes, for they are not only effective in causing back-radiation but also act as condensation nuclei. Smoke-free zones in the London area were first established in 1957, and since then the notorious 'pea soup' fogs have been less frequent; and smogs, thick with soot and sulphur dioxide, like that of December 1952, which caused thousands of extra deaths in the London area, have virtually been eliminated. Central London tended to have denser fogs than the outer suburbs and the surrounding countryside, despite their lower air temperatures and higher humidity; but this is no longer so.

Like other large built-up areas in moist temperate climates, London tends to be cloudier than the surrounding rural areas, and summer thunderstorms develop and intensify to an unusual extent in the northern suburbs, where ridges of higher ground favour the creation of thermals.

Snowfall is also affected by large urban areas, and even in the excessive snowfall periods of 1962–63 there was a tendency for central London to receive sleet rather than snow. Snow also clears quickly from warm streets, but lies longer in the parks, even though surface pollution decreases the albedo.

It must be emphasised that **the actual climatic modifications caused by urban conditions vary from city to city**, and effects which are deemed beneficial in some may be undesirable in others. In cold climates increases in temperature mean greater comfort and smaller fuel bills; in other places additional heat is a burden. Heat generated by concrete and tarmac surfaces, and trapped in enclosed areas, has led some cities in the USA and Germany to introduce more greenery, and to plant trees among large expanses, such as car parks. However, this relates more to the micro-climatic conditions within the canopy layer.

12.4 Urban micro-climates

In the inner cities tall buildings and narrow streets produce patterns of **urban canyons**. This restricts the sky-view and reduces the emission of long-wave radiation from buildings to the atmosphere above.

The properties of the air within the canyons shows **diurnal variations**, much affected by what is illuminated at a particular time and the angle of incidence of the incoming energy during the day. For a north–south running canyon, the upper part of the east-facing wall is irradiated first, and light from the low sun falls onto the face at a high angle. The lower parts, the street, and west-facing wall remain shaded. Though the rest of the wall is gradually lit, the angle of incidence of the sun on the wall decreases; until, towards midday, when it is nearer the zenith, it strongly heats the street. Gradually the west wall is illuminated and part of the radiation is reflected back to the opposite buildings, which retain some of their heat.

During the day about a quarter of the heat received is stored in materials, and the rest is dissipated into the air, creating turbulence, and is removed by convection out of the canyon. This air is warmer than that at roof level and so promotes instability. At night heat retained in the materials is released, and helps to offset radiation losses.

In canyons of a different width and in those running east–west the effects will differ, and, of course, weather conditions affect the radiation exchanges and therefore mechanical and thermal turbulence. So it is obvious that micro-climatic conditions can vary considerably in adjacent parts of the city.

Fig. 12.6 Tall buildings interrupt the airflow and create problems at ground-level. Here in Toronto careful planning has spaced clusters of tall buildings outward through the suburbs, partly to prevent excessive commuter concentrations. But in

Fig. 12.7 **A** Vortices established to the lee of buildings set up persistent eddying. **B** Smoke from a low chimney is sucked into the eddies which develop to the lee of a tall building and swirl among the lower buildings.

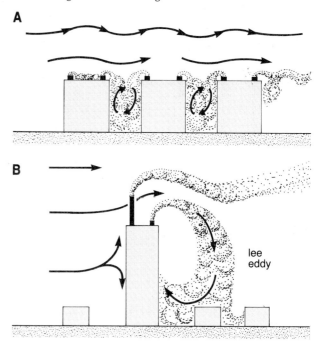

the financial–commercial centre very tall buildings rise above those of an earlier generation, exposed to a wide-open fetch across lake Ontario to the south.

Notice that the tall white building stands on a two-storey podium, and that the entrances to the black buildings are shielded by wide canopies. The frontages of the smaller high-rise buildings are filled in or screened.

There are green spaces throughout the breadth of the city.

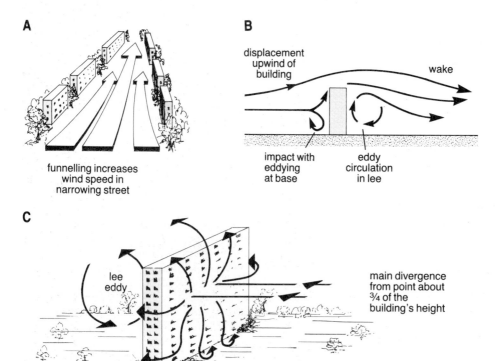

A

funnelling increases
wind speed in
narrowing street

B

displacement
upwind of
building

wake

impact with
eddying
at base

eddy
circulation
in lee

C

lee
eddy

main divergence
from point about
¾ of the
building's height

Fig. 12.8 **A** The Venturi effect; **B** disturbances both upwind and downwind of a tall building; **C** wind striking a tall building can create swirling eddies at street level on the windward side.

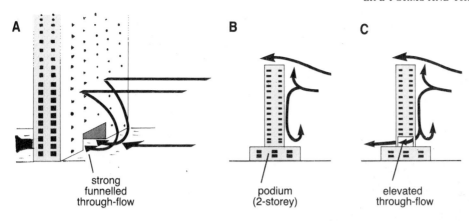

Fig. 12.9 **A** The downward eddying adds to the Venturi effect, increasing the wind speed through a gap in the buildings. **B** and **C** shows ways of lessening eddy swirls at ground-level.

Winds may blanket urban canyon heat exchanges, yet lead to unpleasant, and even dangerous, conditions in certain urban locations. Fig. 12.7 shows how **buildings influence air flows** and can lead to ground-level air pollution. In **A** air is flowing above narrow streets, and even among low housing causes particle pollution to circulate in eddies between the buildings. In **B** a tall building creates a downwash effect in the lee, so that emissions from low chimneys are taken down to ground level. Again, the concentrations of pollutants will depend on whether meterological conditions favour air turbulence or subsidence.

Apart from pollution, the impact of strong air currents on tall buildings, and the flow between buildings and through narrow streets can create unpleasant effects. Fig. 12.8 illustrates **the Venturi effect of increasing air speed** as it flows through narrowing channels, and also the nature of the deflections and circulations about tall buildings.

Fig. 12.9 **A** shows that through-flow tends to create very high speed winds, and such openings, leading to shopping malls, are often important pedestrian passageways. Even the swirling downflows at the base of a tall building whip up dust and litter and make potential shoppers unwilling to linger. In some cases the effects have been lessened by placing the tall building on a podium of one or two storeys high, so that much of the down-flow swirls about the podium roof (Fig. 12.6). In some cases, to lessen the wind force on the podium roof, an elevated through-flow way is set above this, with the main tower block continuing above. Developers are not always able to construct smaller buildings, but they are at least paying attention to wind-tunnels and aerodynamics.

12.5 Urban plant and animal life

Cities are seldom unrelieved areas of brick, concrete and road surfaces; though the heart of the inner city may approximate to an urban concrete jungle, and, as Fig. 12.11 shows, in some cities closely built-up districts extend outward with little obvious vegetation. Most, however, include

parks and ponds and gardens. Trees line the streets, helping to absorb urban noise and, to some extent, break down gaseous pollutants. Some, like the London plane trees, adapt better than other species, many of which succumb to air pollution, or lack of root-space, or deprivation of nutrients in paved areas.

Gardens and allotments occupy surprisingly large areas in the outer suburbs, maintained by inputs of energy, water, nutrients, and pesticides.

Fig. 12.10 Despite the lake-side setting, winters in Toronto are extremely cold, and one response is to create complexes like this, which allow comparison shopping along enclosed walkways at different levels.

In some urban areas, notably in the huge Chinese cities, the intensive cultivation of vegetables in fields and family plots intermingles with dense suburban housing, often on the relics of village land that was enclosed by urban expansion.

In all of these green areas, introduced, carefully controlled plant species achieve temporary ecological balance with animal life and soil organisms. But the city as a whole, even the high-rise centre, has **a perennial wild-life population**.

Fig. 12.11 Athens city and suburbs, where buildings, streets and paved surfaces form a heat trap during the summer months. During the exceptional heatwave of 1987, hundreds died from heat stress and exhaustion. Subsidence helps to form eye-burning *nefos*, a cloud of factory effluents, exhaust gases, and sulphurous fumes from low-grade air-conditioner fuels.

Birds such as sparrows and pigeons, and black-birds in the gardens, form a permanent population linked to human activities – as do the mainly

Fig. 12.12 An urban park where snow lies deep for many days after it has disappeared from roads and buildings. It forms an enclosed system, visited in this case by gulls, herons, and hawks, as well as by local territorial birds, and has an animal life which includes moles, squirrels, and even foxes.

nocturnal, scavenging rodents, who live within the building structures, and the cockroaches and insects. New food chains develop, when towns are established in a wild-life environment (Fig. 10.28).

In the temperate lands, starlings roost on ledges in the warmer inner city, and flocks migrate daily to the outer suburbs and countryside. To most living things the walls, urban canyons, streets, parks, and gardens are extensions of the natural cliffs, valleys, and fields. Animals adapt to urban conditions and human activities. Squirrels, foxes, and even badgers establish territory there, and each country has its own intrusive urban species, like America's skunks and chipmunks. They add to the animal biomass, mostly made up of people and domestic animals.

Plants, with their varied dispersal mechanisms, quickly colonise urban sites – from the lichens on roof tiles to the variety of herbs which root and compete in neglected spaces.

Urban refuse tips also provide abundant nutrition for a wide range of scavengers – from gulls, rats, flies, and cockroaches to bacteria.

13

MONITORING THE SYSTEMS

13.1 Viewing the whole: examining the components

The surface of the earth and the atmosphere are now being scanned almost continuously from space by a large number of satellites. Data relating to selected phenomena is received and processed at short intervals. It is possible for research groups and students to obtain much of this information directly or indirectly; as in the colour plates 20 and 21, which show up-to-date images of a hemisphere or large region with details which are otherwise unavailable, or which would take years to acquire through ground surveys.

Viewing is one thing, of course, but understanding the information provided by satellite data or a false colour image, and appreciating its limitations is another. Most **satellite information** needs to be calibrated from data acquired by practical work in the field, or by other atmospheric measurements. It is also essential to know, for instance, what kind of energy reflections or emissions are being sensed; and therefore which phenomena are being specifically examined, and which will be excluded.

In the context of examining environmental systems, satellite data can be used to monitor progressive changes. We can also use it to observe the distribution of environmental components and the relationships between them; and it enables us to place local studies in perspective within the broader environment. As Chapter 12 shows, for instance, a large urban area is made up of numerous micro-systems; but it also acts as a single unit, modifying the climate downwind, or acting as a source of pollutants. This may well be relevant for a small group engaged on urban–industrial studies; but they are unlikely to be able to gauge for themselves the extent of modification or pollution. They may now, however, turn to organisations with access to technological techniques, such as colour-coded satellite images.

It is necessary, therefore, to be able to interpret information provided by such forms of advanced technology, which can now acquire, through sensors in aircraft and satellites, specific details of cloud formations, heat sources, leaf coverage, soil conditions, urban spread, desertification, etc.

Interpretation of information is particularly necessary when studying interrelationships within a broad region, such as the Italian Campania (p.171). Advanced technologies may be valuable for planning purposes. They may again help to put regional studies in perspective, providing background information, without taking away from the value, and necessity, of carrying out local surveys or field studies of micro-habitats.

In this chapter we look not just at the interpretation of processed data in image form, but at ways in which scanning techniques can be used, relatively inexpensively, in local surveys. There are other ways in which students can make direct use of satellite information and process data for themselves. It is now possible for schools to acquire packaged equipment which allows them to make direct contact with satellites transmitting meteorological data, and so build-up weather images of their own.

From an environmental point of view there are immense advantages in being able to scan the earth's surface and atmosphere with sensors of visible and infra-red radiation, or with radar, or photographic and TV cameras. Apart from adding to our knowledge of the conditions and behaviour of the various components of an ecosystem, these methods are able to give early warning of potentially dangerous situations: in the short-term of developing cyclones, locust swarms, or forest fires; in the long-term of progressive occurrences, such as those leading to desertification.

We need, of course, to **integrate such information with ground-level observations** and measurements in order to substantiate facts and add detail, and also to calibrate data received from the satellite. But it may then be possible to apply this knowledge to satellite images of other areas. This is illustrated by the studies of vegetation described on pp.174–5.

We can also look for information from aircraft or satellites to back up extensive ground surveys; for instance, those monitoring soil deterioration in the Sahel have been able to acquire valuable information from space observations of the frequency and intensity of dust storms.

13.2   Images from space

Satellites in orbit

Satellites with very specific individual functions are put into an appropriate orbit to enable them to obtain and transmit particular types of information. They have to operate at a height which minimises molecular collisions, and so avoids atmospheric drag. There they orbit at the chosen altitude with a velocity which allows the resulting centrifugal force to balance the earth's gravitational pull.

The time it takes a satellite to complete a single revolution about the earth increases with the height of its orbit above the surface. At a speed of about 7.9 km s^{-1} it would orbit the equator, skimming close to the surface, in 84.4 minutes. The higher the orbit, the lower the necessary speed of the satellite, and the longer it takes to complete an orbit (its period of revolution). In the case of a **geosynchronous satellite**, a speed of 3.35 km s^{-1}, enables it to circle the equator in a period of 23 hours 56 minutes (the sidereal day) so that it appears stationary in the sky.

As the earth rotates beneath a lower satellite as it orbits, successive passes of the satellite are displaced westward. The Landsat 4 and 5 satellites scan surface conditions and land use from an altitude of 705 km. They have a near-polar orbit and cover a given area every 16 days or so. They observe the earth's surface directly between latitudes 81°N and 81°S.

The French SPOT satellite, which makes observations for similar purposes from an altitude of 832 km, also has a near-polar orbit, so that its ground-tracks over a 24-hour period appear as in Fig. 13.1. It vertically scans a particular area every 26 days.

One great advantage of the repeat coverages at given intervals is that it is possible to monitor environmental change. For instance, each sensor on the SPOT satellite can view two adjacent strips, covering a swath width of 117 km as it orbits, with a 3 km overlap. It is also possible to steer its mirrors to view obliquely up to 27° either side of the vertical. This increases its coverage to a strip 950 km wide, and thus also increases the frequency of the repeat cycle to a matter of a few days. It also allows stereoscopic imagery, and so makes it possible to determine ground elevations.

Information received

The satellite scanners receive a picture element (**pixel**) as their basic unit. For the Landsats this means reflections from a cell 30 m square. The SPOT operates with a 20 m cell in panchromatic mode, which covers green, red, and infra-red wavelengths; or it can receive data from a 10 m cell when it records the red and green visible light as one band.

Fig. 13.1 The ground tracks of the SPOT satellite with a near-polar orbit, over a 24-hour period. Data received at any point can be stored and transmitted to receiving stations in North America, Western Europe, North Africa, and Australia.

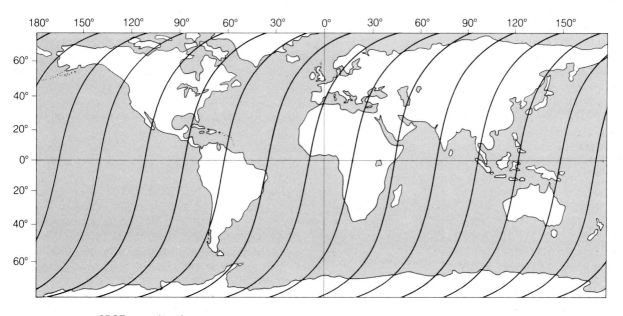

SPOT ground tracks

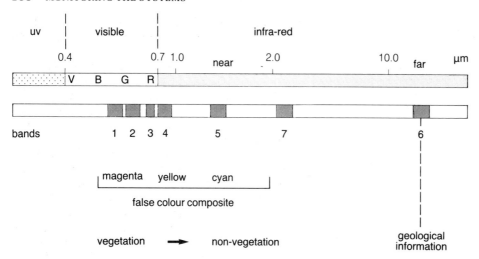

Fig. 13.2   The energy spectrum from ultra-violet to the far infra-red, showing the ranges of wavelengths in bands picked up by the sensors on Landsat satellites.

As it scans the lines of successive strips, **a satellite builds-up information from agglomerations of individual pixels**. The multi-spectral sensors record information from each cell for certain parts of the electromagnetic spectrum. Fig. 13.2 shows the bands of radiation recorded by the Landsats, through the sensors of an instrument called **the thematic mapper** (TM). It senses reflected radiation within the visible spectrum, the near infra-red, and in the longer, thermal, infra-red wavelengths.

Each spectral band is associated with the strong reflectance of radiation from particular components of the ecosystem. Information about specific phenomena – crop characteristics, ocean bed features, or geological structures – can best be found, therefore, from observing the reflectance in a particular band.

Table 13.1 shows that the nature of the reflection from each component gives an identifiable **spectral signature**. Most of the geological information is necessarily received in the far infra-red band; snow, cloud, and soil moisture properties are best examined and differentiated in band 5; and vegetation properties can be distinguished in bands of visible light or near infra-red, according to the state of growth, time of year, and information required; while plant heat stress would be examined from data from a far infra-red band.

Fig. 13.3 **A** shows a generalised spectral signature for vegetation in full leaf, with peak reflectance in the near infra-red. **Different types of vegetation** may be recognised by the characteristics of their own spectral signatures. These will vary with seasonal periods of growth, leaf fall, etc.

Fig. 13.3 **B** also shows that it is possible in this way to distinguish between areas of **wet and dry soil**; and that different types of soil will show strong or weak reflectance of certain wavelengths of energy, which give their spectral signatures a recognisable appearance.

Typical signature curves of reflectance from **various types of rock** can also be used for identification (Fig. 13.3 **C**). Or the indication of strong reflectance of certain wavelengths can be used to concentrate study on those bands which could indicate the presence or absence of certain rocks.

With such a variety of information available, receiving stations in various countries establish links with particular satellites, and either process their data or pass it on to interested organisations. The satellites can either transmit directly to the receiving station, or store information to be transmitted when they are within range, during the appropriate part of their orbit. A number of **global positioning satellites** enable, say, a Landsat to establish its position and direct its imaging apparatus. **A tracking and data relay satellite** (TDRS),

Table 13.1

Band	μm		application
1	0.45–0.52		differentiating: soil/vegetation deciduous/coniferous
2	0.52–0.60	visible	vigour of vegetation (green reflectance)
3	0.64–0.69		chlorophyll absorption shows plant species
4	0.76–0.90	near infra-red	water bodies; determining biomass content
5	1.55–1.75		clouds/snow; vegetation/soil moisture content
7	2.08–2.35	infra-red	identifying rock types
6	10.40–12.50	far infra-red	thermal mapping; plant heat stress; soil moisture

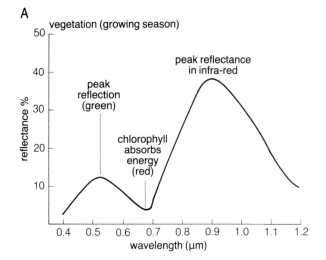

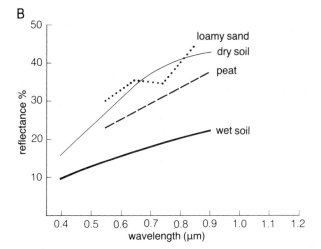

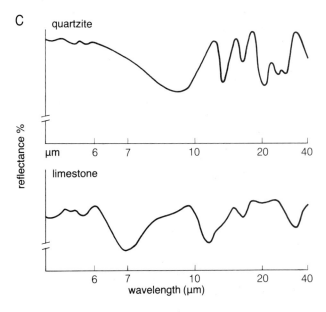

Fig. 13.3 Spectral signatures[9]: the percentage reflection of energy in various wavelengths from (A) green vegetation (generalised); (B) soil types; (C) rock formations.

positioned above, can receive data beamed up from the Landsat and transmit it to an earth station.

Specialist satellites have specific roles. The first of the new series of European remote sensing satellites, ESR-1, will survey, in particular, the oceans and sea-ice, and provide meteorological data from over the open seas, where in the past forecasting has had its limitations.

The **television orbital infra-red satellites** (TIROS), among generations of **weather satellites**, have transmitted visible and infra-red images of weather phenomena. Now known as National Oceanic and Atmospheric Administration (NOAA) satellites, they give a view of both the night and day sides of earth and atmosphere twice a day. The information received is distributed through ground networks for decoding and mapping. Besides these, there is **a global network of geostationary meteorological satellites** (p.75).

The sensors of such weather satellites not only receive and transmit information concerning the lower atmosphere, such as the water vapour content, but also receive data relating to the atmospheric ozone content and temperatures within the stratosphere.

Data from such satellites can also be processed to give images of sea-surface temperatures. These, as we have seen (p.46), are now considered to have great significance for long-term forecasting, in view of their correlations with the strength or otherwise of major atmospheric circulations.

Preparing and processing colour masters

The Landsat bands can be processed in monochrome with 256 levels of brightness on the grey scale; and information from each band may be combined in a number of ways to provide false colour, or simulated natural colour, composites. These are valuable in emphasising particular features, especially those in the infra-red which, otherwise, we would be unable to identify. Plate 20 has been built-up as a composite involving Landsat 5 data from three bands, 3, 4, and 5, giving an **image in false colour** which highlights vegetation through reds, magenta, and pink and non-vegetation features with cyan and blue.

The precision with which such detail can be built up, with fine adjustments of colour and registration, is due to the use of lasers. They enable the data to be transferred to a colour master in a single process. Digital data on tape is converted by lasers to red, blue and green wavelengths, which are modulated, according to strengths, and blended by a mirror system. This automatically focuses them and builds up strips as a colour master on a rotating drum.

The building blocks are still the pixels. As **ground photography or radiography are used in the field to calibrate data from a satellite**, it is often necessary to compare the photographs with digital imagery. Thus a reverse process is also available, which digitises the photographs or maps, and finally records the data on digital tape.

Uses for underwater observations

A false colour composite is also useful for highlighting recorded features concerning **sediments suspended below a water surface**. Remote sensing can be used, of course, to indicate the paths of particle pollutants.

The particles cause more sunlight to be back-scattered than the water does, so the apparent colour of the water varies with the size, composition, and density of sediments. It is also affected by dissolved and organic matter; though, in general, the brightness of the light received by the satellite sensors relates directly to the concentration of sediment suspended in the top few metres of water.

Here, again, fieldwork is needed to calibrate the data; so that water samples are collected at the time the satellite passes over.

Underwater contours and sedimentary deposits can also be revealed by examining specifically the reflectance in the green part of the spectrum. Whereas in the near infra-red band light is strongly absorbed by water, which thus appears black on an image of data from the Landsat 7 band, the green wavelengths in band 4 are not so readily absorbed. The greater transparency allows light to be reflected from a sandy bed down to some 12 fathoms. The deeper the water the more light is absorbed, and so for this wavelength a computer can construct lines of equal brightness to represent bathymetric contours.

The uses of radar

NASA's Space Shuttle has been able to use imaging radar to provide unique information about many areas unsuitable for examination by most of the remote sensing satellites, particularly in those regions which are almost continuously covered by cloud.

Aircraft-mounted systems are also used, and were in operation well before the Shuttle's coverages in the early 1980s. A number of radar satellites are also about to be used for microwave remote sensing.

The techniques and capabilities of radar observations are quite different from those obtained by the remote sensing of wavelengths of reflected natural light. The radar instruments **emit pulses of microwave energy** and **measure the back-scattered radiation** from features on the earth's surface, or just beneath it. This is known as **an active system**. It can, of course, be operated by night or day.

The back-scattering depends on the nature of the topography, the granular composition of surface and near-surface features, the soil moisture content, and the vegetation cover. On the whole, with high angle viewing, coarse materials return the brightest radar images. Lower angle viewing can be used to reveal details of the topography.

The fact that radar can **penetrate dry surface material** gives it many practical advantages, though soil moisture may limit the extent of its penetration. It has been used to reveal fossil drainage systems now covered by a mantle of later deposits, and can thus give information about climatic conditions in the past. Fold and fault structures are also well revealed by radar images.

It has also been particularly useful in **penetrating the canopy of dense vegetation** in the rainforests, allowing studies of the surface and subsurface features.

Airborne thematic mapping

Aircraft are used as support for satellite programmes and are also hired for local survey purposes; and, apart from carrying radar, may have equipment for sensing or photographing reflected radiation. They can fly visible and infra-red sensors under a cloud cover over a given area, at a suitable height, time, and direction relative to the sun.

Some **airborne multi-spectral scanners** may record simultaneously ten or more channels in narrow wavelength bands, from the visible to infra-red. Fixed-wing aircraft, helicopters, or balloons can be specially adapted to a user's needs, and over a narrow swath achieve a resolution of an order of magnitude better than satellite imagery.

Airborne thermal infra-red sensing is widely used for examining the thermal characteristics of soil surfaces, which diffusely reflect heat. There are drawbacks, in that these characteristics vary with moisture and texture, and with near-surface atmospheric influences, such as wind speed, humidity and evapo-transpiration, which modify the thermal infra-red values.

The relationships between remotely sensed red and infra-red radiation and the amount of vegetation make for a much more reliable means of investigation by airborne equipment. Plate 11 shows a false colour image of various components of a heath vegetation, and the practical methods of obtaining a reliable green-leaf area index (GLAI) by a combination of airborne and field radiometry are discussed on p.175. Here the aerial photography provided reflectance data collected by a small format hand-held camera, in order to obtain a perpendicular vegetation index for different types of vegetation. Although a number of flights were needed over a period of time, aircraft hire and simple equipment minimised the expense of the study as a whole.

Plate 13 shows how the airborne thematic mapping data has been used to create **a thermal image, coded to show excessive, expensive heat loss**. It gives a general picture of the differences between the night-time heat radiation from the factory as a whole and the exterior ground surfaces. Even the roads show up as warmer than the adjacent land.

Plate 12 The close cover and small leaves of the heather shade the soil beneath. Bracken develops where old plants, like those in the foreground, leave space and light, and also in clearings, where the peaty soil is better drained. It can rapidly colonise open areas with acidic soils. The vertically received reflectance from a mixed vegetation of heather and bracken changes through the seasons.

Plate 13 The airborne thematic mapper thermal image of a car assembly plant at night. The image is colour coded to show radiant temperatures, and the pink-white areas show a high heat loss, which can be very expensive.

Plate 14 The ashy slopes of the caldera about Monte Somma, colonised by shrubs like the yellow broom and by small brightly coloured herbs.

Progressive colonisation of areas of recent volcanic ash are shown by Fig. 9.11 and in Plates 14 and 15. The development of humus and a new balance of soil nutrients allow higher forms to develop. As the green leaf area increases, so this colonisation becomes apparent on images such as Plate 20, related to the sensing of visible and infra-red wavelengths. Reflections from the combination of woodland and crops seen in Plate 16, contrast with those from the lake surface, and thus appear in contrasting colours in Plate 20.

Plate 15 Dense woods established on the slopes of Vesuvius, bordering the solidified lava flow of 1944.

Plate 16 Fruit trees, vines, and the remains of close woodland almost completely cover the slopes above the crater lake, Lago d'Averno, in the Phlegrean Fields. Beyond are the outlines of Procida and Ischia.

Plate 17 In Campania resettlement farms and intensive cultivation have replaced formerly neglected land, producing a rich vegetation which shows up clearly in the false colour images.

The various forms of land-use in Campania are distinguished on the false colour image, due to their different strengths of reflection in the wavelengths recorded.

Plate 18 The image of these tree-covered slopes on the coastland west of Sorrento, with the lower vineyards, can be seen in Plate 20.

Plate 19 In Campania buffalo were pastured about the moist estuaries of the Volturno and Calore before the resettlement projects were developed. Now most of these reclaimed areas support herds of cattle. The reflectance response from these low pastures differ noticeably from those given by intensively cultivated land, as in Plate 17. Notice the fishing net suspended over the water.

Image processed by the NRSC, Farnborough.

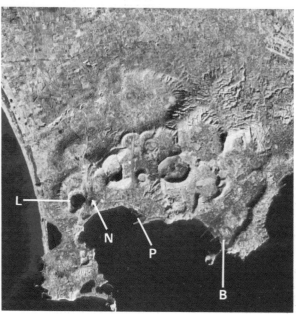

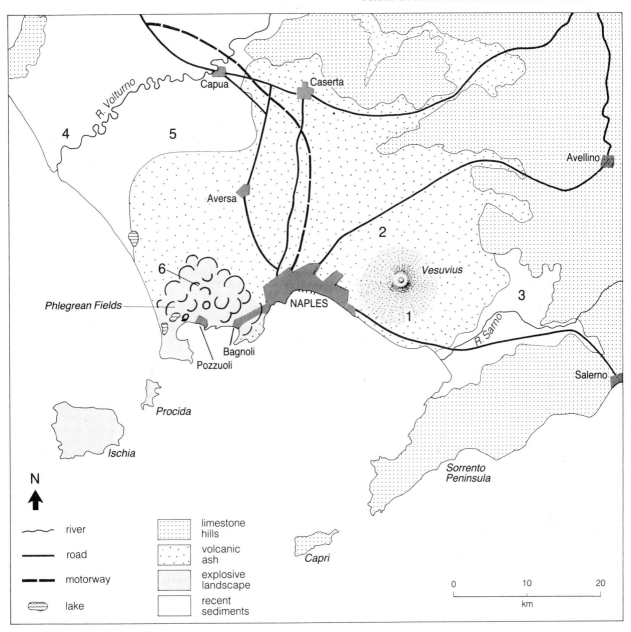

river

road

motorway

lake

limestone hills

volcanic ash

explosive landscape

recent sediments

Fig. 13.4   Most of this northern part of Campania (named after the ancient town of Capua) was below sea-level into the Quaternary, and so most of its surface deposits are recent. It contains immensely fertile plains. Soils derived from volcanic ash cover the whole of the central lowland, and alluvium lies over most of the Volturno valley in the north and the Sarno plain further south.

The false colour in Plate 20 shows vegetation in the range of reds, magenta, and pinks. 1–5 refers to areas where plant coverage, either natural vegetation or crops, is particularly significant.

Non-vegetation features are in cyan and blue, and, of course, include (a) volcanic deposits, both high on Vesuvius and exposed over certain parts of the landscape; and (b) urban areas which are largely devoid of vegetation.

Elsewhere variations in colour tones point to different types of vegetation, and mixtures of crops and exposed soils; seen at a scale of 1:500 000, these make a patchwork of contrasting forms of land use. Compare Plates 20 and 21.

The selected main roads help to focus attention on urban developments, which can be clearly seen in Plate 20.

Plate 20   (opposite above) A Landsat image of bands 3, 4, and 5, showing Campania on 24 January 1983, at a scale approximately 1:500 000. Naples, Vesuvius, and the Sorrento peninsula can be clearly identified. Fig. 13.4 summarises the structure and the distribution of surface deposits, and shows significant routeways.

Plate 21   (below, left) gives a more detailed picture of settlement in the Volturno lowlands; and **Plate 22 (below, right)** shows the Phlegrean Fields, identifying Pozzuoli (P), Bagnoli (B), Monte Nuovo (N) and Lago d'Averno (L).

A more sophisticated satellite-borne heat capacity mapping radiometer was operated from a NASA satellite on a two-year programme. This also measured the effects of loss from urban heat centres, and monitored pollution; though generally the thermal infra-red sensing data was used to locate mineral sources, estimate snow melt-water potential, and for plant canopy measurements.

13.3 Imagery and the examination of complex ecosystems

The satellite image of Campania in Plate 20 gives an overall view of a region where the relations between physical conditions, rural and urban settlement are extremely complex. Here there is physical instability, with active vulanicity and tectonic movements, but a rich soil potential, a climate favouring all-year-round plant growth, age-old settlement, modern industrialisation, new re-settlement schemes, and a huge over-populated city-port.

Yet this false colour image is able to highlight the main features of the structure and soil distribution, simplify the closely intermingled forms of settlement into visible land-use patterns, contrast rural and urban areas of development, and at the same time provide sufficient detail for comparison of cultivation in various sub-regions. It enables us to examine variations in farming practices over quite small areas.

Plates 21 and 22 show how even this degree of enlargement can pinpoint local characteristics and anomalies worth studying in even more detail. An image like this uses reflectance data from only three Landsat bands for the composite, Further valuable information can be obtained from other forms of space sensing over a complex region such as this. Thermal sensing satellites can monitor heat radiation from the very unstable Phlegrean Fields. Other wavelengths provide information of underwater particle movements; not just those visibly transported along the lengthy strip of coastal deposition, but pollutants from the very inadequate sewage disposal outfalls from this large urban area.

A study of the vegetation distribution, indicated by the red, pink, magenta, and orange, will show the value of carefully modulated false colour imagery, and will also provide many examples of the climate–plant–soil relationships which are the main theme of this book. We shall look first at the evidence of vegetation in the areas numbered 1–6 in Fig. 13.4, supported by the colour plates of natural vegetation and land use.

1 Looking first at Vesuvius. Plates 14 and 15 show successions of vegetation, from lichens and mosses to higher plant forms, developing on the almost bare upper cone and the ashy slopes of the caldera. Below, on the lower slopes, the fine ash is enriched by the downwash of minerals, developing soils which are extremely fertile.

On the south of the volcano trees cover the upper slopes (Plate 15). Below a zone of chestnuts there are intensively cultivated orchards, vineyards, orange groves, and vegetable gardens.

2 Inland of Vesuvius, and stretching away northward towards Caserta and westward towards Aversa, is one of the most intensively cultivated parts of Italy. Here many crops are grown side by side; a **cultura promiscua** which includes wheat, maize, vegetables, vines, and orchards of peaches, plums, pears, and almonds. There are often four levels of cultivation together – tall walnuts, small fruit trees, lines of tall vines, and vegetables beneath. The mild climate and adequate rainfall allow a plant cover through the year, its extent clearly shown by the great spreads of red coloration in Plate 20.

Here the dense agricultural population lives in numerous small farmhouses and villages, with clusters in large agricultural towns, such as Aversa.

3 The lowlands about the Sarno form another highly fertile area where a mixture of alluvium and particles of volcanic origin supports the intensive cultivation of deep soils. There are spring-line villages along the lower slopes bodering the plain to the south, and irrigation produces a continuous supply of vegetables.

All these areas of intensive cultivation show up quite clearly on the false colour image, though within these sub-regions there are considerable areas not primarily covered with vegetation. Most of these fall into explicable patterns related to nucleated settlements and the paths of the major roads. There is an absolute contrast, of course, between these mainly agricultural landscapes and the overcrowded city of Naples, where upwards of three million people live in densely populated suburbs, spreading mainly along the coastland. The 1980 earthquake damage to the old city has encouraged a further transfer of inner-city families to peripheral districts.

Focusing again on the rural areas, we can see that to the north the colour tones give a somewhat different appearance and, as we can see in Plate 17, there are more precise arrangements of cultivated land.

4 The lands about the lower Volturno had become sparsely populated over the centuries. Like so many coastal lowlands in southern Italy, the marshlands became malarial, and so neglected. But since World War II, under the direction of the Cassa per il Mezzogiorno, the organisation for the development of the South, much of the land has been reclaimed. Streams have been canalised, marshes drained, new water courses dug, and various types of planned settlement introduced.

In the lower Volturno a regular pattern of pastures for cattle and buffalo has been developed, with new farm buildings. Reflectance from the high proportion of treeless pastures, with grass and fodder crops, shows up as tones of colour which contrast with those from areas which have been converted to productive arable farmland. They appear markedly different from the cultivated areas described in **1–3**, and also contrast with the colours showing land use associated with the older established farms further inland (**5**).

5 The broad valley lowlands south of Capua have long been closely cultivated. In this part of Campania rectangular holdings, dating back to the centuriation pattern of cultivation of the ancient Romans, are still imprinted in the landscape. In some parts large old farms have been broken up. Here land redistribution and the formation of co-operatives have produced a more intensive land use. The satellite image reveals the contrasts between cultivated areas in region 5 and those about the lower Volturno.

6 The most surprising agricultural region in many respects is that of the **Phlegrean Fields**. This volcanic area is a mass of small and large craters, and here and there solfatara emit sulphurous fumes. There are crater lakes like Lago d'Averno, and volcanic tuffs exposed on vertical faces. Thick layers of igneous rock once poured out over this area, which included Ischia and much that is now below the Bay of Naples. A period of great explosions left the greater part below waters and the rest a landscape of overlapping craters. Yet the fertility of the soils is such that vegetation took hold, the area became closely wooded, and people settled and cleared any suitable land for fruits, vines and vegetables. Plate 18 shows how at present a completely green cover of trees and crops runs down to the water's edge of Lago d'Averno.

As on the slopes of Vesuvius, settlement and cultivation are carried on in the face of the long-term expectancy of further volcanic eruptions. In the 16th century the Phlegrean Fields heaved upwards and Monte Nuovo erupted. Today upward movements suggest that magma is accumulating again, and have forced a temporary evacuation of parts of Pozzuoli. Fig. 13.5 shows mollusc holes on the Temple of Serapis at Pozzuoli, which are evidence of former vertical movements.

Plate 21 shows that, although the blue-green coloration indicates exposed volcanic soils, the red areas, extending up the walls of the larger craters, indicate the presence of woodland and close cultivation. In fact, the brighter blue areas in the south-east represent suburban settlement which has been linked to central Naples by tunnels through the outer hills, and also the heavy industrial area about Bagnoli (Plate 22).

About the lowlands **other areas of vegetation** appear in false colour **among the dissected limestone hills**. Incidentally, the orbit allowed sensing to occur at a time of low solar elevation to emphasise the relief features, so that even the erosion gulleys on Vesuvius show up well.

Parts of the Sorrento peninsula are closely wooded (Plate 18), with terraced cultivation on steep slopes; this appears most clearly where beech and chestnuts cover the shaded hills about the town of Sorrento – seen in blue, facing into the bay. Areas of woodland can also be detected among the hills about Avellino and in the enclosed basins to the north-west.

Plate 20 clearly shows **the port of Naples and the inner city area**. Reflection from surfaces of the close-packed tenement buildings, streets, and plazas produce a solid blue appearance when imaged. It also indicates the urban-industrial

Fig. 13.5 The columns of the Temple of Serapis at Pozzuoli, which dominated the market place when the town was a flourishing Roman port. Since then the unstable earth has caused it to sink below sea-level, so that barnacles clung to the stonework. The extent of inundation is shown by the effects of boring molluscs high on the columns. It was heaved up again during the 15th century; and since then the sea-level has fluctuated, with very strong upheavals in recent years.

Fig. 13.6 Steelworks, heavy industries, and suburban extensions, linked to Naples inner city by tunnel, have replaced woodland and cultivation in the area between the volcanic rim and the bay of Pozzuoli. The satellite image records these urban extensions. There are problems through atmospheric pollution and discharges into the bay.

spread south-eastward along the bay, and the developments which closely follow the road through the gap in the hills towards Salerno and southern Campania.

The areas of blue among the cultivated land in the rural hinterland are mostly due to the **towns and larger villages**. Their distribution can be related to the pattern of roads, especially long established routeways, like those indicated in Fig. 13.4. The colour image itself reveals networks of roads within the most closely cultivated districts.

This brief summary shows how clearly the colour images built up by satellite sensing in these particular wavelength bands can be related to the great variety of physical conditions and land use in these northern and central parts of Campania. This is a comprehensive view of a large region; but remote sensing can be of practical value on a local scale. It can again provide data for conversion to false colour composite images; or be used to compare spectral signatures of surface features. Fig. 13.7 shows how graphical presentations can reveal the distribution and state of the natural vegetation and crops.

13.4   Radiometry and local studies

Satellite data acquired at different times of the year can be used for **crop mapping on a field-by-field basis** over large areas, as an alternative to time-consuming surveys. In fact a single image in Landsat 0.50–0.70 μm (green and red) bands is all that is needed to reveal a crop such as rape during flowering time; for its yellow pigment gives it a high radiation response, especially in the 0.50–0.60 μm band (Fig. 13.7A). Other crops can also be identified by examining maximum and minimum data in the various wavelength bands; though it is necessary to correlate these with data obtained during field tests.

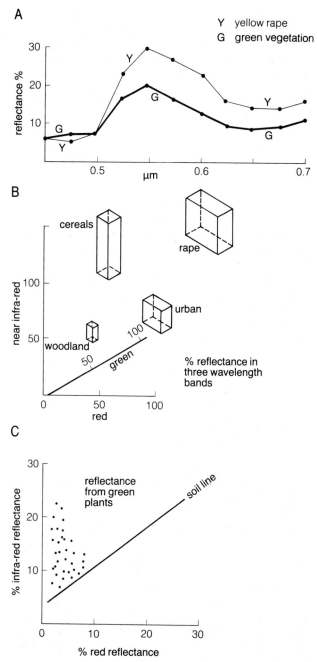

Fig. 13.7   A Yellow rape is distinguished by its high reflectance close to 0.55 μm; this decreases in the red, for chlorophyll absorbs this energy wavelength for photosynthesis; it then increases again in the infra-red.[10]

B Reflectance in green, red, and infra-red wavelengths plotted to produce characteristic shapes and locations of boxes.

C The result of green plants producing a higher infra-red reflectance and lower red reflectance than the bare soil.

Fig. 13.7B also shows that a **three-dimensional plot** of the spectral response in three different wavelength bands can provide a typical 'box' for various surface phenomena, including specific crops, woodland and urban areas. It is thus possible to identify forms of land use spatially, and obtain agricultural census information without the immense labour involved in organising field surveys and correlating their information.

Nevertheless, drawing on satellite surveillance and commissioning surveys by remote sensing apparatus carried in aircraft prove expensive ways of providing data for local academic studies. Thus much work is being done with the aid of **hand-held radiometers and infra-red photography**; and this can be extremely useful provided the necessary errors involved are appreciated and allowed for. It is worth taking note of some of the practical considerations.

The amount of light reflected by natural vegetation or a crop canopy in any wavelength band depends on the age and state of plant growth, the health of particular plants, and the spatial distribution of the individual plants.

When the reflectance values for bare soil in the red and infra-red are plotted against each other, a straight line results – **the soil line** in Fig. 13.7C. As plants grow and produce a vegetation canopy the infra-red reflection increases and the red reflection decreases. Thus when readings of the spectral response of the plants are plotted, their position on the graph tends to move away from the soil line, perpendicularly. So the perpendicular distance of the points from the soil line relate directly to the plant cover.

The condition of the plants can also be revealed by the fact that, for instance, a nitrogen deficiency is shown up by much lower values in the red/infra-red ratio. This can be spotted well before the unhealthy state causes a visible change in the crop itself.

In each of the above, readings can be obtained by means of a portable field radiometer, which records reflectance from 0.25–2.50 μm in seven wavelength bands. It can also be mounted on a tractor, or in a light aircraft.

Plate 11 shows an infra-red picture of a heath area which is burnt over from time to time. It was taken from a light aircraft during a series of observations to test the accuracy of **calculating the green leaf area index (GLAI) by field radiometry**: that is the proportion of the plan area of green leaves to a given land area.

The GLAI varies seasonally, with the maturity and health of the plants, and with the changing dominance of particular species. In a heath area, for instance, it can be used to investigate the rate at which bracken is replacing heath plants. There are relatively cheap ways of investigating this, but the results must be acceptably accurate. The methods used to test this tell us much about climate–soil–plant relationships, and the studies of this heath area are worth summarising:

1 Four vegetation associations were selected: young ling; mature ling; bracken; and bracken and ling. The bracken would give much greater GLAI values.

2 In each, five points (0.2 m²) were randomly selected, and on 12 dates during the year a number of radiometer and photographic measurements were made from a height of about 3 m (from a step ladder), and of adjacent bare ground. Solar elevation affects reflectance in the various wavelengths differently. The effect was minimised by choosing cloudy days.

3 At each sample point vegetation was harvested from three 0.02 m² quadrats, and all the leaves were spread out and photographed. By projecting the transparencies onto a screen with random sample points, the number of green contacts with these points could be counted. This enabled the proportion of the transparency areas covered with leaves to be found and the GLAI to be calculated.

4 The soil line was determined for each site and the vegetation responses for each vegetation association observed, as in Fig. 13.7C. The reflection received from the vegetation canopy varies with soil background reflectance, so its mean perpendicular distance from the soil line was used as a **Perpendicular Vegetation Index (PVI)**. Regressions against the GLAI showed their relationships for each of the plant associations.

5 Near-vertical 35 mm air photographs were obtained from hand-held cameras in light aircraft: one with infra-red film, the other with the same film/filter combination for multispectral photography as on the ground. On 20 days areas of the four associations (6 m²) were photographed (together with ground targets of known reflectance as a check). The reflectance data were transformed to PVI values, and GLAIs calculated.

6 Within the sample areas three sample points were again harvested, and the PVI and GLAI determined.

The accuracy of the method was found to be acceptable; though in fact it varied with the season. In summer the ling flowers hid many of its leaves; in spring ling plants hid many young bracken leaves; while the bracken area was little affected. Reflectance from old litter also varies according to seasonal growth.

These then are some of the different levels at which spectral scanning can assist a wide range of practical planning. Even at school level, data is now available from remote sensing, and field studies of the kind outlined above are within the scope of student groups.

Portable radiometers such as the Spectrascan are available, recording reflections in seven special bands in the wavelength range 0.25–25 μm. For predictions of crop yields or plant stress, they can be mounted on a tractor for readings from a particular field.

CONCLUSION

These are but a few examples of methods we can now use to observe and monitor processes affecting the climate, soils, vegetation, and the results of our occupation of the earth's surface. The earlier chapters aim to give an insight into these processes and the interactions between the major components of the global ecosystem. We need an informed view not just of our own surroundings, but of the deteriorating environments and rapidly dwindling resources in lands which should be supporting many millions of people, but are failing to do so. An understanding of these processes and an ability to interpret information now provided in abundance by advanced technology ought at least to pin-point precautions and suggest practical remedies.

PART 2 GLOSSARY

aerosol: a mass of very small solid or liquid particles suspended in the atmosphere

agronomic: relating to the management of agriculture

algae: plants containing chlorophyll but without differentiation into root, stem and leaf and adapted for life in water or a damp environment

anaerobic: conditions devoid of free oxygen

analogue model: a means of predicting climate from a past situation with similar features to that being considered

bathymetric: referring to measuring the depths

biological magnification: a process whereby consumers in a food chain acquire a higher concentration of a chemical substance at each trophic level

boundary layer: the layer of air adjacent to a surface

chelation: the process of incorporating a metal ion directly into the molecular structure of a compound

cyan: a dark blue colour

defoliant: a substance or process which strips leaves from plants

dystrophication: causing degeneration of plant structure

eutrophication: raising the net primary production of an organism by adding nutrients

fungi: plants without chlorophyll, obtaining food from other plants, animals and decaying organic matter

halogen: one of the four elements – fluorine, chlorine, bromine and iodine – which have closely related chemical properties

isotopes: atoms of the same element but differing in atomic weight; nearly all elements in nature are a mixture of several isotopes

laminar: referring to a flat, thin layer

panchromatic: sensitive to light of all colours

permafrost: permanently frozen sub-soil

photochemical reaction: the interaction of chemicals caused by exposure to radiant energy

phytoplankton: microscopic aquatic plants; the primary producers in the ocean

pixel: a square surface area whose reflections form the basic unit received by satellite scanners

podium: a projecting raised base or pedestal

sensible heat: heat that can be sensed (ie with a thermometer)

soil line (substrate line): a straight line resulting from plotting red against infra-red reflectance from a bare soil site: used as a basis for comparing similar wavelengths of reflectance from types of vegetation growing in that soil

spectral band: a narrow sequence of wavelengths of emitted or reflected energy

tephra: particles blown from a volcanic vent

urban canopy layer: part of the atmosphere affected by micro-scale processes below roof level

Venturi effect: the increased velocity of a fluid flowing through a constriction, resulting from the need to conserve mass

wind chill: heat loss from a body as wind removes sensible heat and causes latent heat loss through evaporation

GENERAL INDEX

PLACE NAME INDEX